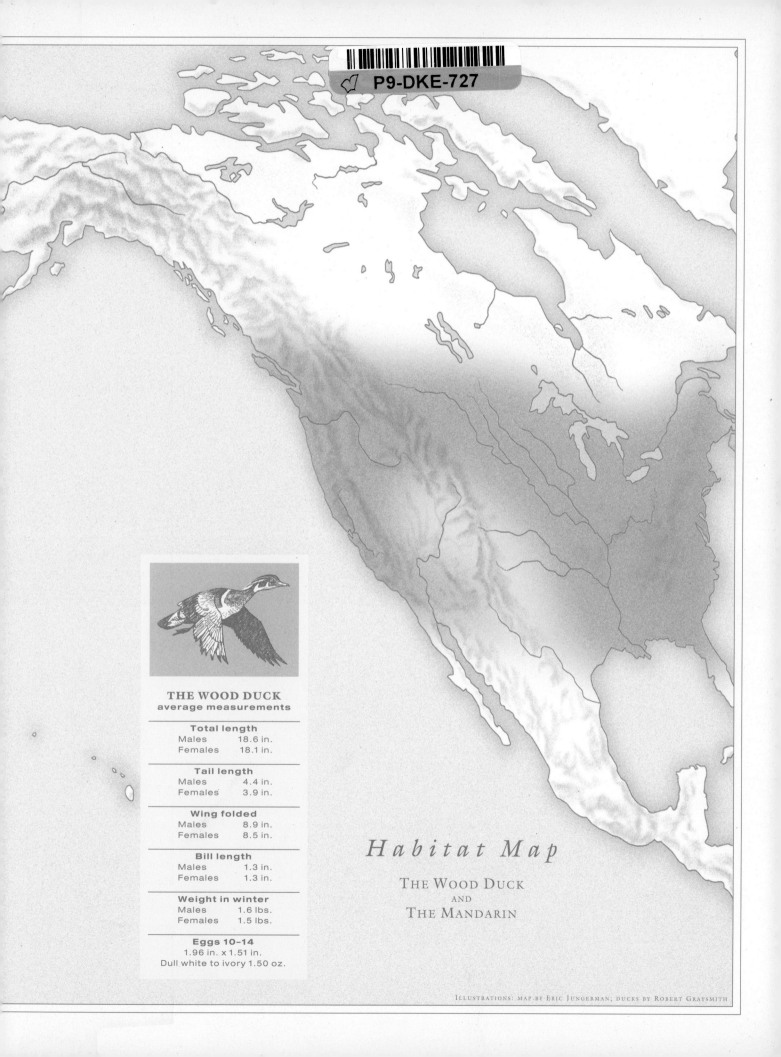

P9-DKE-727

THE WOOD DUCK
average measurements

Total length
Males   18.6 in.
Females   18.1 in.

Tail length
Males   4.4 in.
Females   3.9 in.

Wing folded
Males   8.9 in.
Females   8.5 in.

Bill length
Males   1.3 in.
Females   1.3 in.

Weight in winter
Males   1.6 lbs.
Females   1.5 lbs.

Eggs 10–14
1.96 in. x 1.51 in.
Dull white to ivory 1.50 oz.

Habitat Map

THE WOOD DUCK
AND
THE MANDARIN

ILLUSTRATIONS: MAP BY ERIC JUNGERMAN; DUCKS BY ROBERT GRAYSMITH

THE
WOOD DUCK
AND THE
MANDARIN

Genus *Aix*

"The only two
Wood Ducks nesting
in the temperate regions,
the North American Carolina*
and the East Asian Mandarin,
are probably the most
beautiful members of the
whole family. They are
typical Wood Ducks
frequenting forested
regions and nesting
in tree-holes."

DR. JEAN DELACOUR
"The Waterfowl of the World"
Vol. III, 1959

*Carolina is the European name for the
North American Wood Duck

THE
WOOD DUCK
AND THE
MANDARIN

THE NORTHERN WOOD DUCKS

BY

LAWTON L. SHURTLEFF

AND

CHRISTOPHER SAVAGE

UNIVERSITY OF CALIFORNIA PRESS

Berkeley   Los Angeles   London

University of California Press
Berkeley and Los Angeles

University of California Press, Ltd.
London, England

Copyright © 1996 by Lawton L.
Shurtleff and Christopher Savage

Cataloging-in-Publication data is on
file with the Library of Congress.
ISBN 0-520-20812-9

Printed and bound in Hong Kong by
C & C Offset
9 8 7 6 5 4 3 2 1

EDITOR: Judith Dunham
PROOFREADER: Desne Border
INDEXER: Karen Hollister
COVER AND INTERIOR DESIGN:
    Visual Strategies, San Francisco
MAPS: Eric Jungerman
ILLUSTRATIONS: Sarne *(chapter heads)*
    Robert Graysmith *(appendix)*

Photo Credits and Permissions

Location of photograph on page is described by the following abbreviations: t=top, m=middle, b=bottom, l=left, r=right. Vertical columns of photographs are numbered from the top (c-1, c-2, etc.).

Front cover: Wood Duck (t) by Carl R. Sams II, Mandarin (b) by Scott Nielsen; back cover: Wood Duck (t) by Steve and Dave Maslowski, Mandarin (b) by Carl R. Sams II.
George Ackerman: 25 (t, m); 49 (t); 52; 66 (t, b), 67 (t, b); 97; 98 (c-1–4); 114; 122; 127; 171 (bl); 211 (b); Tupper Ansel Blake: 32 (tr); 69 (t); Del Bogart: 184 (t, b); Hugh Clark: 164 (t); Sharon Cummings/DPA: 207 (t); Ray Cunningham: 104 (b); Andy Davies: 160; 161; 168; 171 (t, br); Jack Dermid: 89 (c-2–6); 92 (t, m); 96 (br); 101 (m); 104 (t); 105 (t, b), 187; Larry Gates: 63 (t); Seton Gordon: 163; Goto: 150 (t); Mike Haramis: 96 (tr); F. Eugene Hester: 99; 188 (t, b); Joe MacHudspeth: 20; 57; 63 (b); 72; 78; 82; 95 (t, b); 108; 109 (b); 110; 111 (b); Barry Hughes: 120 (b); 192 (t, b); 198 (t, b); 202; 203 (t, bl); 204 (t, b); Mark Hulme/WWT: 207 (b); Toshiaki Ida: 148–49; 154–55; 158–59; 159 (tr, mr, br); 212–13; Ichiro Kikuta: 150 (b); 153 (tr); 156; Stephen Kirkpatrick: 101 (t); Shotaro Kosaka: 153 (tl, br); S. J. Lang/VIREO: 93 (b); Steve and Dave Maslowski: 58; 84; Greg Nelson: 28 (tl); 32 (ml); 36; 125; 166; Scott Nielsen: 22 (t); 33; 64 (t, b); 65 (l); 68 (b); 69 (m, b); 86; 87; 88 (t, b); 90–91; 92 (b); 94 (t); Kenneth Rice: 56; 189 (b); 191 (t); 194 (t); 195 (m, b); 196 (m, b); 203 (br); 210; Kelvin W. Sampson: 62 (tl, tr); Carl R. Sams II: 19; 21; 23 (t); 24; 25 (b); 28 (ml, r); 32 (tl); 35; 42 (b); 46; 47 (t, b); 48; 51 (tr, br); 53; 60 (b); 68 (t); 81; 93 (t); 101 (ml); 113; 123 (t, b); 128 (l); 132 (b); 132–33; 136–37; 139; 142; 144–45; 146; 162 (t, b); 164 (b); 165; 170; 172; 173; 189 (t); 190 (t, b); 191 (b); 197; 211 (t); Christopher Savage: 148 (tl); 149 (tr); 152; Lady Scott: 200–01; Yuri B. Shibnev: 119; 120 (t); 121; 128 (r); 129 (l, r); 134; 137 (r); 205; Anneke Shurtleff: 34 (c-1); 195 (t); Lawton L. Shurtleff: 14; 16 (l); 16–17; 22 (bl, br); 23 (bl, br); 26; 27 (t, b); 29; 30 (bl, tl, m, tr, br); 31; 32 (bl); 34 (c-2–4); 37; 38 (t, bl, br); 40 (l); 40–41; 42 (t); 43; 44; 49 (b); 50–51; 59 (b); 60 (t); 65 (r); 70; 74; 76; 77; 79; 89 (c-1); 90 (b); 94 (b); 96 (l); 100 (t, m, b); 101 (mr); 102 (t, b); 105 (m); 106; 107; 109 (t); 111 (t); 124; 126; 130; 131; 138; 140; 167 (t, m, b); 193; 196 (t); 206; 214; 216; 218 (tl, bl, r); 218 (r); 219 (tl, tm, tr, b); 221; 222 (l, r); 223 (t, b); Jean F. Stoick: 28 (bl); 59 (t); Larry Stone, courtesy Shawn Kaufman: 194 (b); William J. Weber: 101 (b); Anonymous: 141.

Wood Duck by Basil Ede (191) is reproduced courtesy of The Warner Collection of Gulf States Paper Corporation, Tuscaloosa, Alabama.

Contents

Forewords 7

Introduction 9

CHAPTER ONE

The Northern Wood Ducks
Together in the Wild 15

CHAPTER TWO

The Abundance, Decline, and
Recovery of the Wood Duck 55

CHAPTER THREE

The Home and Habits of the Wood Duck 83

CHAPTER FOUR

The World of the Mandarin 115
  Russian Ussuriland 119
  China 134
  Japan 146
  Great Britain 160

CHAPTER FIVE

The Mandarin in Art and Literature 175

CHAPTER SIX

Conservation 187

Appendices
  I Nestbox Program for Wood Ducks or Mandarin 214
  II Banding Mandarin in the Far East 221
Resources
  Bibliography 224
  Organizations 226
Index 227
Acknowledgments 232

FOREWORDS

Since I wrote a foreword to Christopher Savage's book on the Mandarin Duck in 1952, the species has continued to flourish as an introduction to Britain, and the Wood Duck of North America has recovered from the threat of extinction earlier this century, but continues to depend heavily on wise management. The Mandarin Duck in the Far East, without management but with official protection in China, Japan, and Russia, appears to have declined to a fraction of its numbers forty years ago.

Lawton Shurtleff and Kit Savage have combined their efforts to record the dramatic stories as well as their own personal observations of these two closely related species. Kit has brought up-to-date his work on the Mandarin with the benefit of five years of residence in Japan, and Lawton has had the enviable and remarkable experience of nurturing a wild colony of "self-introduced" feral Mandarin on his ranch in northern California alongside a natural population of Wood Ducks.

The two authors aspire to achieve the preservation of these two species in the wilds of their native habitat for the joy of future generations of mankind. This is by no means a forlorn hope, as the authors show. Their magnificent color photography enables the reader to share the spectacular beauty and lifestyle of these two birds. They have also brought together a wonderful collection of cultural and biological material on the Mandarin and the Wood Duck, which can be of great value in conservation education. The concept that we are dependent on the natural world for our own survival is the beginning of all wisdom.

It can be maintained that any and every species of animal should be conserved in a dangerous and destructive era, but somehow the Mandarin Duck and North American Wood Duck are so exquisitely beautiful that their survival is likely to become a high conservation priority.

As the twentieth century draws to its close, the twenty-first has to become the Century of Global Conservation. These two beautiful ducks could be a kind of talisman for that.

My late and great friend and fellow waterfowl addict, Jean Delacour, wrote, "The only two Wood Ducks nesting in the temperate regions, the

North American Carolina [as the North American Wood Duck is known in Europe] and the East Asian Mandarin, are probably the most beautiful members of the whole family." In my opinion he should have left out the word *probably*. No two birds better deserve a book to themselves than the Wood Duck and the Mandarin.

— SIR PETER SCOTT
Slimbridge, England

This book about the Wood Duck and the Mandarin is, in some large measure, a creation of Sir Peter Scott, my late dear friend and fellow wildlife artist. It is filled with the colorful images he so hoped for and is beautifully written—all in a fine book about two of the world's most beautiful waterfowl and a book he would have been proud of.

— ROGER TORY PETERSON
Old Lyme, Connecticut,
United States

It is an honor for me to add my words to those of Sir Peter Scott as a foreword to this book. I first met Peter Scott when my friend Bristol Foster and I dropped into Slimbridge in 1957 at the beginning of our "'Round the World" Land Rover trip. As the years passed, I was privileged to enjoy the friendship of a fellow wildlife painter and, of course, one of the most remarkable and important men of the twentieth century.

I could not improve on his eloquent words concerning *The Wood Duck and the Mandarin*. He expresses my feelings completely about the gentle grace and luster of these two birds. I have painted both species and, of course, all eyes (especially artists') revel in the glory of the males' plumage. However, the subtlety and form of the females display a classic elegance which suggests the wild and vulnerable wooded wetlands of this world.

— ROBERT BATEMAN
Fulford Harbour,
British Columbia,
Canada

FOR THE PAST HALF CENTURY, history has been made in the world of waterfowl in the Pacific Flyway of North America, though it has gone largely unrecognized. In a small area in northern California, the North American Wood Duck has been joined, little by little, by another species of wood duck, and a close relative, the exotic Mandarin, a native primarily of China, Japan, and other areas of East Asia bordering the western rim of the Pacific Ocean. On the family ranch in Sonoma County in northern California, we have watched with great interest and curiosity the ever-increasing number of spectacular Mandarin that have come to visit us. The bright copper color of their unique sail feathers and the brilliant white of their breasts brought flashes of light to the dark waters of the lake beside our ranch home.

By the mid-1980s, several hundred Mandarin had established themselves on our ranch and close-by areas as a true feral, and perhaps self-sustaining, population that was breeding, nesting, raising their young, and flying wild alongside the native Wood Ducks of the Pacific Flyway. Never before had Mandarin been known to nest in the wild on this continent. These two magnificent birds from opposite ends of the world—the North American Wood Duck, *Aix sponsa*, and the Mandarin, *Aix galericulata*—are the only species in the genus *Aix*. Of the seven species of wood ducks worldwide, they are the only ones that nest in the northern hemisphere and are therefore the only ones referred to by the appropriate name of Northern Wood Ducks.

Having enjoyed the experience of observing these two splendid birds, I was fortunate in 1983 to meet Dr. A. Starker Leopold, a distinguished professor of biology at the University of California at Berkeley and one of the few people on the West Coast knowledgeable about the North American Wood Duck. Starker Leopold was a son of Aldo Leopold, the world-famous conservationist, and the nephew of Aldo's youngest brother, Frederic, a well-known authority on the Wood Duck, who had studied the ducks for almost fifty years in the Mississippi Flyway. Since we were engaged on our California ranch in a program of setting out man-made nestboxes for Wood Ducks and since

9

Starker Leopold had a similar program for Wood Ducks on the Sacramento River in the north Central Valley of California, the two of us had much in common to discuss—and I had many questions to ask.

When I described to him the unusual presence of Mandarin flying side by side with Wood Ducks—and even nesting in nestboxes designed for them—he reacted with great interest and questioned me in detail: Where did the Mandarin come from? How long had they been there? How many did there seem to be? How did the two species interact? Did they ever interbreed? Was either species dominant? He seemed to sense the existence of a unique situation and, ever the inquisitive biologist and teacher, urged that I consider undertaking a serious study to learn more about the natural history and particularly the interrelationship of these two Northern Wood Ducks.

We discussed at length the desirability of having an introduced or exotic species of waterfowl in the Pacific Flyway. Leopold expressed interest in the subject of introduced, or naturalized, species—the rare successes and the more frequent failures. He described the popularity of the Chinese Ring-necked Pheasant, one of the few imported wildfowl to have prospered in this continent. With great success, it has replaced many of those native prairie-chickens and Sharp-tailed Grouse unable to adapt to the cultivated croplands that had supplanted their prairie habitat. He seemed to feel that if the Mandarin, like the pheasant, had found what he described as a "vacant ecological niche" in North America, it could be an equally attractive and welcome addition to the wildlife population. Also, it would be the first time that any alien species of ducks or geese had ever established itself on the North American continent. This was my first clue that we might be witnessing avian history in the making.

However, Leopold was quick to add a stern note of caution: that, on principle, he was opposed to the introduction of alien species unless they contributed significantly to the environment without threatening the existence of indigenous species. In this case, the species of greatest concern to him was the highly esteemed North American Wood Duck, which has very special and often limited requirements necessary for its survival. On the other hand, if a duck as unusual and distinctive as the handsome Mandarin did not threaten the Wood Duck's status or debase its singular beauty by crossbreeding, it could be an exciting addition to the Pacific Flyway. Careful records should be kept of its migratory patterns, if any should develop.

We questioned why no one had written about two such closely related and fascinating ducks as the North American Wood Duck and the East Asian Mandarin. It was Leopold's opinion that we in the West knew far too little about the Mandarin and that the rest of the world knew too little about the Wood Duck. If Mandarin were finding a new home alongside the Wood Duck in the Pacific Flyway of North America, the time was ripe to learn more about each of them. I recall his parting remark—that the heart of any such book on these beautiful birds should include outstanding color photography to illustrate their splendor and reveal their distinctive lifestyle. His enthusiasm for elegant photography was undoubtedly prompted by the latest of his many wildlife publications, *Wild California,* on which he was then working and which ultimately contained magnificent full-page color images by Tupper Ansel Blake. Sadly, this was my first and only meeting with Leopold, who passed away only a few months later. There have been many times in the years that followed when I would have sought his help and counsel.

So, with the guarded encouragement of this respected biologist, teacher, and conservationist, we began a serious study of the Wood Duck and the Mandarin throughout the seasons on the streams and ponds and in the oaklands of our ranch in northern California. Equally pleasurable and instructive was the opportunity to share experiences and ideas with other experts in the field of waterfowl. As the result of my encounter with Starker Leopold, I met his uncle Frederic Leopold, Dr. Frank Bellrose, S. Dillon Ripley, Henry M. Reeves, and Arthur Hawkins, and in England students of the Mandarin including Sir Peter Scott, Sir Christopher Lever, Dr. Janet Kear, Andy Davies, and lastly Christopher Savage.

Meeting Christopher Savage in 1984 was especially propitious. He had been contemplating an update of his 1952 book, *The Mandarin Duck,* which was inspired by the prosperous colony of Mandarin in England, a land as foreign to the Mandarin as North America. Fascinated by my experience with the Wood Duck and the Mandarin in California, he was enthusiastic about writing a book about the Northern Wood Ducks. What was more natural than our collaboration on such a book, the first ever devoted to an in-depth study and natural history of these two greatly admired and closely related waterfowl.

Thus began more than ten years of intensive research. Christopher, by coincidence, moved to Japan, where he was able to study the Mandarin in its native habitat. In addition, he met with most of the surprisingly few serious students of the Mandarin in Russia, China, South Korea, and, of course, Japan. For the first time he was to hear and read eyewitness accounts, almost all of which described the Mandarin as a seriously endangered species. For a bird once so highly esteemed in Asian cultures, this seemingly unchallenged threat to its existence was a paradox that was difficult to understand. Christopher's earlier experience with the Mandarin in Great Britain and his new insight from Japan convinced him that we should undertake a serious effort—almost a crusade—to make the world aware of the Mandarin's history and its beauty, and to urge help in its conservation. Just as the North American Wood Duck needed advocates early in the century when the species was faced with possible extinction, the Mandarin now needs the same kind of protection. Conservation efforts beginning in the early 1900s at all levels—by federal, state, and local agencies, with great assistance from the private sector—had saved the Wood Duck from extinction in North America. Much could be learned from these successful programs that might help save the endangered Mandarin.

As Christopher began to explore the largely unchartered territory in the rapidly vanishing world of the Mandarin, I crisscrossed North America to meet with authorities on the Wood Duck and to observe its special habitats. From the great Butte Sink of central California, to the quiet rivers and lakes of New England, to the blackwater swamps of the Deep South, the Wood Duck was obviously thriving. More successfully than almost any other species of duck, the North American Wood Duck is adapting to an ever-changing environment and finding new breeding areas or rediscovering old ones long abandoned.

In the Pacific Flyway, where the Wood Duck and Mandarin fly together in the wild, the Wood Duck population is believed to be increasing faster than in any of the other flyways on the continent. Although the Mandarin could conceivably make history by finding a new home on the Pacific Flyway, its presence, in our judgment, would in no way threaten the Wood Duck. Rather, the Mandarin will prosper only in those occasional areas that offer abundant food and nesting opportunities, and, then, only if assisted—as the introduced colony of Mandarin in Great Britain has been for the past seventy-five years—by the helping hands of landowners, farmers, conservationists, and other lovers of waterfowl who provide them with feed and nesting facilities.

Of greater importance than the Mandarin's potential is that the ducks are here—bringing their elegance and special character to this continent without any evidence of prejudice to its fauna and flora.

The Wood Duck and the Mandarin: The Northern Wood Ducks is making its own history as the first book to celebrate these two very special and closely related species and to describe the Mandarin's presence in North America. Following the natural history of the Wood Duck and the Mandarin through the seasons, the text and photographs take the reader to three very different regions of the world—across North America, to coastal areas in Russia and China, then to Japan, and from there to England. The journey goes back in time to capture the distinctive but equally rich histories of both species and leads forward to anticipate their future. The future is in large part in the hands of dedicated individuals and organizations working worldwide to protect valuable forests and wetlands on which so much wildlife relies. By journey's end, the reader will be able to share with Christopher and me our insight into the workings of a well-organized, but little-understood, web of conservation worldwide and our deep admiration and affection for two of the world's most beautiful waterfowl.

—LAWTON L. SHURTLEFF

CHAPTER ONE

THE NORTHERN WOOD DUCKS TOGETHER IN THE WILD

Watching the wild Wood Ducks and Mandarin on the lakes at your ranch is like witnessing the world premiere of a brilliant ballet. The costumes are the plumage of two of the world's most colorful waterfowl; the set is the water, the woods, and the skies of our California wine country.

—MFK, GUEST BOOK, INDIAN MEADOW RANCH (1982)

O N THE WEST COAST OF THE UNITED STATES, on the thousand-mile-long Pacific Flyway, the North American Wood Duck and its closest relative, the Mandarin of East Asia, fly together in the wild for the first time in history on a true Wood Duck flyway. It is a spectacular sight to watch these brilliantly colored birds on the wing, on the ponds and streams or in the trees, perching, preening, or, in early spring, searching for their special nest sites. Both ducks nest here on Indian Meadow Ranch in northern California, where my family has lived for more than twenty years, giving us the opportunity of observing, studying, and photographing the Wood Duck and the Mandarin through the seasons—courting, breeding, nesting, and raising their young.

Since the early 1700s in Europe and more recently in the United States, these two elegant birds have enthralled and been the pride of aviculturists on private estates, who enjoyed, studied, and, on rare occasions, wrote about them. By contrast, the Wood Duck and Mandarin here at Indian Meadow Ranch are free-flying—neither penned nor pinioned. Native Wood Ducks on

OPPOSITE At Indian Meadow Ranch in northern California, the native North American Wood Duck meets the only other Northern Wood Duck, its closest relative, the golden Mandarin from the Far East.

For the first time in history, Wood Ducks and Mandarin share the forests and wetlands of the Pacific Flyway in North America. Here in northern California, they nest and brood their young in a seemingly identical ritual.

the Pacific Flyway range over an area extending west to the Pacific Ocean, east to the Sierra Nevada, south to central California and occasionally to the Mexican border, and north through Oregon and Washington and well into southwestern Canada.

The several hundred Mandarin resident at Indian Meadow Ranch and adjacent areas in Sonoma County are descendants of feral immigrants far removed from their native habitat. In the process of adapting to an environment so temperate and so distant from their homeland, they have lost most of the migratory instincts of Mandarin in the Far East, which range over an area from southeasternmost Russia down the eastern coastal plains and mountains of the Koreas and China, even to the island of Taiwan. Mandarin farther north cross the Sea of Japan and make their way down the Japanese archipelago to Okinawa, the most southern of the Ryukyus, and west into parts of Myanmar (Burma) and India. Mandarin have been sighted recently in Vietnam, northern Thailand, and Nepal.

In late autumn, the total population of Wood Ducks in North America is estimated at seven to eight million birds. However, scarcely over one hun-

The Pacific Ocean separates the two Northern Wood Ducks. Mandarin in their native flyways range from Ussuriland in southeastern Russia to the countries of Southeast Asia. This territory is similar in latitude to the distribution of Wood Ducks in North America, from the southern provinces of Canada to the states along the Gulf Coast. The usable habitat of the Wood Duck is probably three to five times larger than that of the Mandarin in Asia.

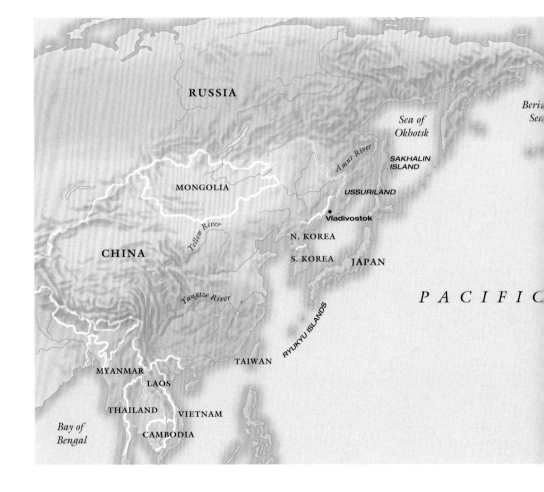

dred thirty thousand inhabit the Pacific Flyway, and, of these, only twenty-five to thirty thousand reside in central and northern California. The other ninety-eight percent of all the world's North American Wood Ducks are found east of the Rocky Mountains on the Mississippi and Atlantic Flyways, including southern Canada.

Indian Meadow Ranch is near the geographic center of the fabled California wine country, fifty to one hundred miles from the center of some of the state's most densely populated Wood Duck wintering and breeding grounds in and around the Butte Sink of the Central Valley. The vast fertile valleys of the wine country are crisscrossed and separated by foothills, giving each valley its vintage

From the cabin at Indian Meadow Ranch, Wood Ducks and Mandarin can be seen through the surrounding oaks as they land on the nearby lake for their evening feeding.

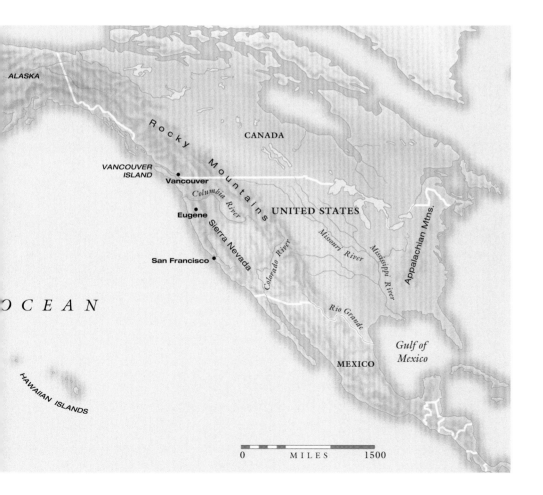

The myriad colors in the plumage of the Wood Duck drake provide perfect camouflage within the shadowed forests and sunlit wetlands of its habitat.

designation—Napa, Sonoma, Alexander, and Mendocino, among many others. For centuries, the Russian River and its innumerable tributaries have provided the region with rich and inviting Wood Duck habitat.

The Wood Duck *(Aix sponsa)* and the Mandarin *(Aix galericulata)* are, in many ways, unique in their relationship. First, they are the only two species in their genus—*Aix*. This is one of those rare instances among ducks where only two species constitute the entire genus—the graceful hens appearing like twin sisters, the handsome drakes like suitors in contrasting colors. While the Wood Duck and the Mandarin share this intimate scientific relationship, they live in habitats that are almost at opposite ends of the earth. Wood Duck and Mandarin drakes, though completely different in appearance, are often described as two of the world's most beautiful ducks, possibly the most beautiful of all waterfowl. Both are scarcely half the size of a Mallard. Yet their magnificent colors make them the cynosure of all observing eyes.

The Wood Duck drake, with his iridescent green head, back, and tail feathers, blends into the background of a forest or the dark waters of a pond. Closer inspection, however, reveals a rainbow of colors: the brilliant red of his

The Mandarin drake's dramatic copper-colored sail feathers are unique among all species of waterfowl.

eyes and beak, the diamonds of white that fleck his russet breast, the even more gleaming white of his throat and sideburns, and the tawny tones of his barred flank feathers, often referred to by fishermen as "flytyer's gold."

By contrast, when the Mandarin drake is on the water, his dominant colors of copper, buff, and white create a seemingly golden image that contrasts sharply with his surroundings. His most conspicuous feature is a pair of unique copper-colored sail feathers, actually part of the tertial plumage, which extend vertically two inches or more above and forward of his tail feathers. His cheeks and neck are the same brilliant copper hue accentuated by a large triangle of pure white above his dark eye. Traces of luminous iridescent blue on his back and sides, even on the sail feathers, highlight the otherwise golden image. His flank feathers, the almost identical barred buff color as those of the Wood Duck, and also long prized by flytyers, suggest the close relation of the two species. When the Mandarin is perched on a bank or a log, even riding high upon the water, his underside shows a much more gleaming white than that of the Wood Duck. Both drakes, when excited, display a gorgeous head crest, the Wood Duck's a deep green, the Mandarin's a blue-green

and copper. There is no confusing the darkly handsome Wood Duck drake and the golden Mandarin.

Perhaps because of the gentle nature and graceful carriage of the Mandarin and Wood Duck hens, they are considered by many the most beautiful of female waterfowl. In their subtle hues of grays and browns, highlighted by touches of white and turquoise in their wing feathers, the two hens can barely be distinguished, one from the other, even by experts. The Mandarin hen may be slightly more grayish than the Wood Duck, but the primary distinguishing feature is that the Mandarin has a finely drawn line of white behind her eyes, whereas the Wood Duck's eyes are encircled by a heavier

At a distance, only the keenest wildlife watcher can distinguish the female Wood Duck, with the graceful white teardrop around the eye, from the female Mandarin.

tracing of white, almost a perfect teardrop. A major difference, although rarely noticed, is that the underfeathers of the Mandarin's wings are a dull blue-gray, while the underlying coverts of the Wood Duck are a colorful flecking of grays, tans, and white. Even less known is that several of the secondary flight feathers on the female Mandarin are splashed on the upper side by randomly placed white stripes rarely found on the Mandarin drake and never on

Wood Ducks. During their summer molt, when Mandarin drakes and hens in their drab plumage are almost indistinguishable, these white splashes, located above the white tips of the secondaries, help to identify the sexes.

Very young Wood Duck and Mandarin ducklings, still in the down, are equally difficult to dis-

tinguish from one another, though the Mandarin shows a hint of saffron yellow, while the slightly smaller Wood Duck is a soft gray. As we watch them scampering over the water in response to the call of their hens, or chasing an errant insect, we think of them as our gold and silver bullets.

In late spring, both drakes begin to shed their bright plumage—first the Mandarin drakes lose their spectacular sail feathers, and a few weeks later the Wood Duck drakes enter their eclipse molt. Soon the ponds are dotted with discarded feathers. As if this loss of their handsome coloring were not enough, within two weeks both species lose all their flight feathers in what ornithologists refer to as the "remigial molt," a phenomenon characteristic of most

ABOVE The eye of the Mandarin hen is encircled by a fine white line that extends to the back of the head and fades behind the neck.

OPPOSITE On the top side of a female Wood Duck's wing, splashes of purple, gold, and blue are visible above the white tips of her secondaries. Alongside are her greenish gold tertials. A rich speckling of gray and tan patterns the underside of her wings.

LEFT The top side of a female Mandarin's wing is less colorful, showing only a speculum of metallic blue, but is distinguished by the several unique white stripes on her secondary wing feathers above their brilliant trailing edges. The underwings are a somber gray.

This rare mixed brood probably resulted from a female Mandarin laying her eggs in a Wood Duck's nest, since the Woodie ducklings outnumber the Mandarin. The downy Wood Ducks are a soft silvery gray, the Mandarin a slightly saffron yellow.

waterfowl. Then, for several weeks, Wood Ducks and Mandarin, unable to fly, are extremely vulnerable to every type of predator. It is a time of crisis for nearly all waterfowl, whose flightless period varies among the species. Yet in all the literature on Wood Ducks and Mandarin, this critical aspect of the molt is never given a specific name. The term *remigial* refers to the remiges, the primary and secondary flight feathers, and the Latin root of the word means "oarsman." What a fitting description for the feathers that propel birds through the sky and the loss of which, as in the case of the Mandarin and Wood Duck, renders them helplessly earthbound.

The Wood Duck and the Mandarin are truly birds of the forest—both species seek out old tree holes in which to nest and lay their large clutches of eggs, and they perch and preen among the branches. In the depths of the forest, they hatch their young and find security in sheltered streams and quiet ponds, avoiding the open water most other ducks prefer. Both species have exceedingly large eyes, long, wide tails, and broad wings that enable them to fly safely through dense forests, their wing tips barely flicking the branches. Wood Ducks and Mandarin take readily to man-made nesthouses that simulate natural tree cavities. The frequent placement of nesthouses close to human habitation makes it easy and exciting to observe the ducks nesting, their eggs pipping and hatching, the day-old ducklings dropping safely from nests often twenty or more feet above the land or water, and finally the watchful hens brooding their chicks until they are able to fly away on their own. To observe the life cycles of these two spectacular species native to opposite ends of the world is a memorable experience, which for our family began more than twenty years ago.

S ince then, on many a cool fall evening on our porch at Indian Meadow Ranch, we have listened and watched with awe as native Wood Ducks spiraled down into the main lake directly below the cabin, surrounded by towering Redwoods. In the moments separating dusk from darkness, with the moon in its cycle rising over the distant hills, the first awareness of the arrival of Wood Ducks came from the explosive sound made by their braking wings—a sound that, once heard, is never forgotten—as they headed for splashdown on the ponds before us. Thus alerted, we could follow the swift, plunging descent of flock after flock still visible against the wooded hills, their silvery breasts barely shining through the darkness.

Once the ducks were on the water, the quiet of evening would come alive with the calls unique to the Wood Duck. The first was the soft, nervous chattering of the ducks as they checked their new surroundings, only to be interrupted by the high-pitched squeal of an alerted hen. A rare instant of silence would be broken by one of the most distinguishing of all Wood Duck sounds, the soft *keer-loo-keer-loo*, the last syllable always rising plaintively—the coquette call of a lovelorn hen summoning her mate or hoping to attract one. Those were—and still are—for us true Wood Duck evenings before the onset of winter. To understand the presence of Wood Ducks in such profusion on the lakes at the ranch, it is helpful to know something about this particular area of California and the Pacific Flyway in general.

Indian Meadow Ranch consists of some two thousand acres of typical foothill country located approximately an hour's drive north of San Francisco and fifty miles inland from the Pacific Ocean. Cutting through the heart of the ranch is Barnes Creek, a raging torrent of water in late fall, throughout winter, and into early spring. Dozens of tributary streams rush through steep canyons to feed the main creek. Annual rainfall ranges from less than twenty inches in the dry years to more than eighty inches in the wettest. From the ranch,

TOP After generations of Mandarin have been raised in captivity in North America, the species has apparently been imprinted to prefer to nest in man-made nestboxes.
ABOVE AND RIGHT Native Wood Ducks prefer natural tree cavities, but will use nestboxes when cavities are scarce.

Barnes Creek passes through several other ranches before entering Mayacama Creek and, finally, the Russian River one and one-half miles to the north—an area that throughout history has attracted Wood Ducks.

Redwood, willow, bay, and alder line Barnes Creek along most of its course through the ranch. The hillsides are densely covered with gnarled, lichen-draped oaks—California Black, Blue, Oregon White, California Scrub, and Coast Live—which produce the acorns that are the staple of the Wood Duck's diet. Madrone, Toyon, manzanita, and California Wild Grape offer fruits and berries to birds and animals alike. On a close-by hilltop stands an Acorn Woodpecker's "granary tree." An ancient Black Oak bleached almost white by the many seasons, it is studded with myriad acorns in all shapes and sizes, stowed in holes drilled in nearly vertical lines, the sharp ends pointing inward. Although dozens of fresh green acorns have been added, almost half the holes stand empty, waiting for the woodpeckers to finish their work in late October. In fall, winter, and even into early spring, acorns are a major staple of the Wood Duck's diet here and in other parts of the country. Thousands of acres of cultivated vineyards in the lowlands surrounding the ranch provide the Wood Ducks a potential bounty of unharvested grapes in late fall and winter.

A Wood Duck drake and his mate float on one of the sparkling lakes at Indian Meadow Ranch in the renowned wine country of northern California.

The gradual disappearance of oak trees on the West Coast is a major threat to the future of Wood Ducks in this region. Competition for the acorn mast is fierce among insects, Band-tailed Pigeons, Wild Turkeys, deer, wild pigs, rodents, and other mammals. Of the many millions of acorns dropped from untold numbers of oaks in the northern California foothills, few survive to send forth a sprout. Of those that do, rabbits, deer, wild pigs, and cattle ensure that only a rare oak seedling ever becomes a full-grown tree. At the same time, the Douglas-fir are moving slowly down from higher ground to take over range formerly occupied by oaks. If the oaks are finally replaced by the fir, the Wood Ducks will have no choice but to move on to more marginal wetlands and forests of deciduous hardwoods to supply their nesting and foraging needs.

The first priority when we arrived at the ranch in 1968 was the construction of a lake among the Redwoods. We wanted it to have an island or two and be the location for a summer cabin to be built above the shoreline. The lake was built at the confluence of Barnes Creek and several lesser streams that provided turbulent winter runoff. While it was under construction, an old-timer, a local operator of a bulldozer on the job, first told us about Wood Ducks in the area. As a young boy, he explained, he and his friends hunted ducks along the creeks on the ranch—but in the last twenty to thirty years all of them seemed to have disappeared. He suggested that putting up some nestboxes when the lake was filled might bring them back. Wood Ducks, he explained, nested in the hollows of old trees. If these were not available, they used nesthouses, or nestboxes, man-made simulations of natural cavities in trees.

He then offered to show us what a Wood Duck nestbox looked like. We walked below the newly constructed dam, and in the space of several hundred yards, hanging approximately twenty feet above the streambed, in alder and bay trees, were three old nestboxes. Using a ladder, we inspected them: boxes approximately one foot in width and two feet deep, with a four-inch-diameter entry hole in the front, about three inches below the roof. They were obviously built to resemble the hollowed-out trunks that a nesting duck could enter through a hole created by a rotted-off branch or chiseled by a large woodpecker. These boxes were old and weathered but, being redwood, were still in excellent condition. All three were filled to overflowing with litter left by Western Gray Squirrels, and there was no evidence of recent use by

In anticipation of winter, colonies of Acorn Woodpeckers store their bounty in an oak snag known as a granary tree—a practice unique to this colorful species. The Wood Duck's fondness for acorns inspired early observers to name it the Acorn Duck.

ABOVE, FROM TOP Columbian Black-tailed Deer, Red-winged Blackbirds, and Belted Kingfishers share the forests and streams of the ranch.

RIGHT Great Blue Herons are resident the year around.

Wood Ducks. So we took down the three old boxes, refurbished them, and constructed six new ones. In early spring of 1969 when the new lake was filled, we set out the nestboxes. Three were placed on wooden posts in the lake near the shoreline, and three were hung twenty feet or so up in the Redwoods close to the water's edge. The other three were installed, one each, on tiny, well-wooded cattle ponds high in the hills far removed from the new lake. To hurry the return of any kind of ducks, we put out decoys and imported and released one pinioned pair each of several species of waterfowl, recorded in the old ranch log as "teal, gadwall, pintail, wood ducks, widgeon, redhead, and scaup." A note written later at the bottom of the page reads "All lost within the year—too many predators?"

Of the more than one hundred nestboxes installed beside the streams and ponds at the ranch, more than half are used each year, almost equally, by Wood Ducks and Mandarin, and most are successful in producing large broods of ducklings.

Whether attracted by the decoys or the imports, in spring the first few wild Wood Ducks did return. However, they left just as quickly as they had arrived. Since the ponds were new, vegetation in and around them was sparse. The shore provided little cover for the shelter that Wood Ducks require. The ducks also might have departed in search of food on the many neighboring ponds where evening feeding by the ranchers, mostly of Mallards, was a regular practice. From that time on, morning and evening, we started to throw a few handfuls of grain along the shoreline of the main lake. The vegetation on which ducks thrive— pondweed, duckweed, water-fern, milfoil, algae, and the ubiquitous cattails—soon began to form.

The next year brought the first two nesting Wood Ducks. What excitement we felt as we watched this great event through field glasses. A Woodie hen hatched her ducklings in one of the nestboxes twenty feet above the main lake and, miraculously, was on the water below, calling them to jump to her side. One by one, they tumbled out, eight tiny balls of fluff bouncing through the branches and finally landing with scarcely a splash beside the anxiously calling hen. A week later, a nearby nest produced a second brood of twelve ducklings, and we felt that spring and the Wood Ducks had really arrived. Mysteriously, however, neither brood was ever seen again. We soon learned what had happened.

A year earlier, a friend had brought to the ranch a pair of large white Muscovy Ducks. The Muscovy nested early in the newly planted shrubbery on the lower island of the lake, hatched a brood of twenty pale yellow ducklings, and,

One year, a Wood Duck hen nested in a natural tree cavity close to the cabin at Indian Meadow Ranch. Thirty days later, twelve precocial downy ducklings were hatched. Some twenty-four hours later, they scrambled to the nesthole and, summoned by their mother's call, tumbled down to join her on the lake below the nest.

being completely domesticated, paraded them around the lake. The earlier mystery of the disappearance of the wild chicks was solved when we saw one of the yellow Muscovy ducklings pulled with a splash beneath the surface of the lake. The heavy Muscovy hen dove with surprising speed to bring the struggling chick to the surface. We retrieved the duckling only to find it so badly injured it could not be saved. Of the original twenty ducklings, only five ultimately survived. The predators, we realized, were surely some of the more than one hundred Largemouth Bass with which we had stocked the lakes, a common ranching custom in this area. A large bass can hold an object the size of a tennis ball in its powerful jaws. To this day, the first Muscovy hatch invariably occurs before the first hatch of the wild ducklings, and has become our predator alert system.

Each year thereafter, we added a few more nestboxes, but even though more and more ducklings hatched, we rarely saw a Wood Duck brood on the water. Innumerable predators, we soon learned, threatened both the hen and her clutch in their nests in the trees. These included Raccoons, Bobcats, Opossums, and occasionally Mink and Ringtails. After the ducklings hatched and were with their hens on the water or ground, foxes and skunks joined the

list of predators. In the air, Cooper's, Red-shouldered, and Sharp-shinned Hawks, Great Horned Owls, and even a pair of Golden Eagles were added perils. In the water, the bass and later giant Bullfrogs took a heavy toll of ducklings—but of all these, bass accounted for the greatest loss. While they were a popular attraction for the anglers among our friends and neighbors, they were disastrous for the ducklings. So we painstakingly moved the larger fish to other ponds and, to satisfy our angling friends, restocked the lake with less predatory fish—Bluegill, Redear Sunfish, and a few catfish.

Bass had obviously been the major problem, for by the fifth year after setting out the first nestboxes, Wood Duck hens and their ducklings were visible on the lake. At last we had a Wood Duck program that was exciting for us and our nature-loving guests to observe. By the late 1970s, we had erected one hundred twenty-five nestboxes and built more than twenty cattle and duck ponds.

The first week of each February, when we inspect more than one hundred nestboxes just prior to the nesting season, is one of the busiest and most interesting times of the year at the ranch. We climb ladders and trees and stumble down hillsides, with all the paraphernalia needed to refurbish the

After the ducklings leave the nest, the Wood Duck hen leads them to shelter and water plentiful in insects, invertebrates, and other high-protein foods. If necessary, she will lead the ducklings over a mile to suitable brooding habitat.

nestboxes for next year. This is also a final check on last spring's hatch as we log in which nestboxes were used, which were successful in hatching duck-lings, and which were unsuccessful, having been abandoned or destroyed by predators—and which were not used at all. We also estimate the number of eggs laid by counting the membranes in the nest and noting any eggs left unhatched.

Since neither Wood Ducks nor Mandarin bring any foreign material to their nests, all new nestboxes are filled with wood shavings to a depth of approximately four inches. Each year we remove and replace the old shavings in boxes where eggs were hatched the previous year. Experience has taught us that bumblebees, working the winter manzanita blossoms whose fragrance fills the hillsides, choose for their hives boxes containing last year's duck litter, rendering those boxes useless until they are cleaned and refilled with fresh shavings. Soft pine shavings seem to attract nesting hens far more than do fir shavings. Sawdust often packs too hard.

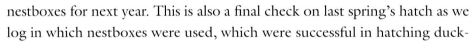

TOP LEFT AND RIGHT On land, Bobcats and Ringtails
take a heavy toll on Wood Duck hens and their broods.
ABOVE In the water, Bullfrogs and bass cause havoc with
young ducklings.
LEFT At the nest, Raccoons are an ever-present menace.
OPPOSITE The Wood Duck's large eyes, broad wings, and
exceedingly long tail enable it to fly through dense forests.

By mid-February, the first hens are actively seeking nest sites. Even with the thousands of oaks on the ranch, we are able to locate only a few natural cavities. Yet some Wood Duck hens obviously do, for they produce chicks that were not hatched in our nestboxes. One natural cavity that we did discover is less than thirty feet directly opposite our kitchen window. In a single season, a hen makes twenty or more trips to and from the nest to lay her eggs and well over one hundred additional trips during incubation of the eggs. In all these many trips each year to and from the nest near the cabin, so secret and stealthy were the hen's movements that we never once noticed her. The nest was discovered by coincidence one afternoon when twelve or fourteen Wood Duck chicks jumped from the nest, only a few yards away from where we were sitting, and tumbled after the hen into the lake directly below us.

In the early spring of 1972, an epochal event occurred at the ranch. We were walking in the woods at the edge of the lake with a frequent visitor, a neighbor and knowledgeable waterfowl fancier, who handed me his binoculars and pointed to a nestbox on one of the lower islands. Clearly visible, even without his glasses, was a startlingly beautiful duck perched in a tree beside the box. I could not remember ever having seen such a bird before. At a distance he seemed all copper and gold.

"That," said our neighbor, "is a Mandarin drake, a close relative of the Wood Duck." He went on to tell us that his old friend and mentor, the late Dr. Jean Delacour, an internationally acclaimed ornithologist, had classified the Wood Duck and the Mandarin as "Northern Wood Ducks" to distinguish them from a few species of larger wood ducks in the southern hemisphere, such as the Australian Wood Duck and the White-winged Wood Duck. The North American Wood Duck, our friend informed us, is known scientifically as *Aix sponsa*, "waterfowl in bridal dress," and the Mandarin as *Aix galericulata*, "duck with a little bonnet." "Many experts," he continued, "believe the handsome Mandarin drake and his elegant hen are the most beautiful of all waterfowl. I'll wager it came here over the hills from my own pond nearly twenty miles away."

He explained how Mandarin often had escaped from his father-in-law's extensive wild bird collection. In fact, he had met his wife when, as a boy, he tended her father's pens. On one occasion, he recalled, a door blew off a pen and an entire flock of young birds escaped. For years he has kept a few pair of Mandarin in pens of his own, and some free-flying Mandarin use a lovely pond in his garden where he maintains six nesthouses. Another small colony of Mandarin nests in boxes set out on Sonoma Creek near his home. Both groups of free-flying Mandarin are located less than twenty miles from Indian Meadow Ranch as the ducks fly.

LEFT The first Mandarin seen at Indian Meadow Ranch was found inspecting a Wood Duck nesthouse. Both the drake and his hen stayed and raised their brood, marking an event that forever changed life at the ranch.

OPPOSITE, FROM TOP The end of the nesting season is the time to check the nestboxes. The membranes of hatched eggs indicate the number of ducklings that hatched. A nest with cold and uncovered eggs and traces of an adult duck's feathers suggests that a predator caught the hen in the nestbox. A nest is occasionally used by squirrels or woodpeckers to store acorns. Wooden nestboxes with a small side door make inspection easier and quicker.

The Mandarin drake can be identified, even in flight, by the gleaming white on his head and breast.

The next morning, with renewed excitement, we saw a Mandarin hen leave the nestbox where we had seen the drake. Already it contained her first three eggs buried in the soft pine shavings. Little did we realize then the impact and pleasure this occasion would ultimately bring to our lives. That was almost twenty years ago.

This October evening, as once again we sit waiting for the sound of ducks flying in to the lake, little has changed. Pileated Woodpeckers emit their piercing cries as they fly toward their roost in the Redwoods. Wild Turkeys gobble in the distance. Red-tailed Hawks scream their sharp *keeeer-keeeer* as they circle high above the fir trees. The Belted Kingfisher ventures out for its last dive, chittering as it lopes through the night air. The Redwoods planted many years ago on the lower island of the main lake have more than quadrupled in size, and the lake is now abundant with foliage.

Suddenly tonight, a small flock flashes in, and the air explodes, as always, with a rush of wind through outstretched wings. This is the haunting sound of Wood Ducks braking, twisting, and turning in preparation for splashdown, the same welcome sound we have heard so often throughout the years. The quiet evening erupts with the excited piping of birds newly arrived behind the upper island and the wistful *keer-loo-keer-loo* of the Woodie hen calling her mate. Only this evening there is a subtle difference. Here and there in the descending flock appear flashes of gold, and clear patches of pure white breast plumage reflect off the water where none so bright was ever seen before. Tonight, also, an unmistakably new note is heard, a rapid, staccato *uib-uib*, short and strident like the bark of a small puppy—the commanding sound of the Mandarin drake carrying the length of the pond. Wood Ducks and Mandarin are everywhere around us.

What a joy to see that, brood by brood, in ever-increasing numbers over the past fifteen years, the Mandarin has joined its cousin, the North American Wood Duck. Very little has been written about these unique, also tree-hole nesting ducks in their native habitat in Russia, China, Japan, and nearby areas of Asia. This is partly because they live in countries that, until recently, have not communicated with each other or even welcomed visitors, and partly because in many of these areas the secretive and shy birds are relatively unknown. Mostly they are found only in small numbers over a vast and often unexplored area. Many knowledgeable observers believe that the Mandarin may be nearing extinction in its native habitat.

Pair by pair, and brood by brood, over a period of twenty years, the exotic Mandarin have established a free-flying population of several hundred birds that remain in Sonoma County on the Pacific Flyway.

We still don't understand exactly how or why, of all the ranches in this area, the Mandarin chose ours. The only plausible explanation for their concentration here is that we have an established feeding program and have constructed a series of wooded ponds and erected scores of man-made nesthouses that are as attractive to Mandarin as they are to Wood Ducks. Neighboring ranches generally do not have nestboxes and have, at the most, a cattle pond or two. Their cattle quickly graze off young alder or willow sprouts, cattails, and other emerging vegetation, leaving little of the cover that is indispensable for Wood Ducks and Mandarin. The main lakes at Indian Meadow Ranch are protected from cattle, so the shorelines and islands offer ample protective shelter to attract both species.

The lake below the ranch cabin provides ideal brooding habitat. By summertime, protein-filled plants cover the water. Succulent duck-weed and fibrous water-ferns are two of the ducks' favored foods.

The flyways of the Wood Duck and the Mandarin, although widely separated geographically, are similar in many ways. The Pacific Flyway, where the ranch is located, and the two eastern flyways of North America range in latitude from approximately thirty to fifty degrees north, whereas the Mandarin's native flyway, five thousand miles west across the Pacific Ocean, ranges from approximately twenty-five to fifty-five degrees north. The latitudes are similar enough to suggest the Mandarin might adapt to the Pacific Flyway of the West Coast. Everything we have learned of the Mandarin here and in its native habitat confirms that its behavior and needs are almost identical to those of the native Wood Ducks on the Pacific Flyway.

So different from all other ducks, yet so intriguingly similar to each other, the Wood Duck and the Mandarin have become a major interest on the ranch. Nothing so brightens our mornings as to waken to the first rays of sunlight glinting through the Redwoods above the pond where dozens of Wood Ducks and Mandarin loaf on the water and along the shoreline. Then, at the slightest movement in the cabin, they are off in a flashing of wings—the piercing alarm call of the Woodie hen interrupting the early quiet. Within moments a few flocks return, circling several times above the treetops before spiraling in to continue to feed and frolic. Other ducks rarely come to the lake. Except for the ever-present Great Blue Heron and occasionally a small Green Heron, in spring a pair or two of Mallards nesting on the islands, in winter a few Common Mergansers, and even a Double-crested Cormorant or American Coot—Wood Ducks and Mandarin have taken almost exclusive possession of our ponds. Whether they shun or are shunned by other ducks is not clear. However, here the two species live together in relative isolation, and their obvious compatibility, constant interplay, and apparent comfort in each other's presence reinforce our awareness of their friendly, fascinating, and close relationship.

An almost symbolic example of this relationship occurred early one spring many years ago and has been repeated nearly every year since. An extremely tall grove of Redwoods stands on the far shore of the main lake in view of the cabin. Early in the morning we noticed an unusual activity too high in the trees to be seen clearly with the naked eye. Binoculars revealed an amazing sight, given the majestic height of the Redwoods and the miniature appearance of several pairs of Mandarin and Wood Ducks fluttering about in the branches well over one hundred feet above the lake. First one hen and then another flew from the branches to the trunks of the Redwoods to peer into holes carved by members of the colony of Pileated Woodpeckers that nest here. While the holes are unusually large, they are too small for a duck to enter or use as a nest. Yet we watched as time after time the hens flew from one hole to another and back to the branches. What attracted them—two species native to different parts of the world—to search for nest sites nearly a hundred feet higher than either has ever been known to nest?

Then we recalled that much of the literature concerning Wood Duck nests has described woodpecker holes, particularly old holes made by Pileated Woodpeckers, as ideal nesting sites. We fantasized that both birds, from continents apart, had read the literature and noted the Pileated Woodpecker holes but had failed to consider that, for whatever reason, the holes in the Redwoods were too small for nesting. The birds learned this lesson rapidly, however, for in a few moments they abandoned their efforts—probably to

Wood Ducks and Mandarin idle together serenely on the lake below the cabin. When startled by the slightest sound or movement along the shore, they rise from the water, their wings flashing in the sunlight. Within moments, they will have disappeared into the forest.

search for the snags of old oaks where the digging was easy.

Shortly after the first Mandarin arrived at the ranch, we observed another impressive bit of evidence of their intimate relationship with the Wood Duck. On a lake some distance from the cabin, we saw for the first time a mixed brood of nine very young ducklings—five Wood Ducks and four Mandarin—led by a Mandarin hen, with a Mandarin drake following close behind. For several days we watched this unusual brood swimming in and out of the cattails and overhanging willow branches—but on the fourth day they were nowhere to be found. Imagine what disbelief and then rejoicing we felt as only a day later the same brood swam into view on the main lake below the cabin. There was no mistaking such an unusual brood, with the Mandarin drake still in attendance. The entire brood, completely intact, had followed the hen overland, across dry fields and rocky

ABOVE During midday, Wood Ducks and Mandarin retreat to the shoreline under the overhanging willows and into the shadows of the towering Redwoods.

RIGHT The gentle call of the California Quail emanates from the forest undergrowth in the early morning as Wood Ducks and Mandarin fly onto the lake to feed.

outcroppings, for more than a mile. Since then we have had the good fortune of observing scores of Mandarin courting, nesting, and hatching their golden ducklings and on rare occasions of watching the most exciting of all events—the hen calling her day-old chicks out of their nest to join her on the water below.

From time to time we have seen other mixed broods of Wood Duck and Mandarin chicks, invariably with a Mandarin hen. But through the years, as more and more nestboxes have been made available, we have seen fewer and fewer mixed broods. A possible explanation, knowing that Wood Ducks, and probably Mandarin, have the keenest sense of smell of any duck, is that a hen of one species senses that a nest is being used by a hen of the other species and leaves to find a nesthouse of her own. In any event, despite continuing to see

Two Mandarin hens escort their mixed broods of ducklings—the silvery Woodies and the golden Mandarin. It's an unusual sight since Mandarin and Wood Ducks rarely lay their eggs in each other's nests.

large broods—twelve to fifteen ducklings—of both Wood Duck and Mandarin, for the past five years we have never again seen a mixed brood.

Nor have we ever seen a hybrid of the two species. Wood Ducks produce hybrids with several other species such as Mallard and teals, but Mandarin do not appear to produce hybrids with other species, including the Wood Duck. One commonly accepted explanation is that the Mandarin has a different chromosome count than other ducks. However, we have observed and studied several cross-matings—inevitably of a Mandarin drake with a female Wood Duck. In only two instances were we certain that eggs resulted from the cross-breeding. Both hens incubated their eggs well in excess of thirty days before abandoning them. Upon examination, all were found to be infertile. Claims of hybrids hatched from a cross of Mandarin and Wood Ducks surface from time to time, but to our knowledge none has ever been substantiated.

Although the many similarities of the Wood Duck and the Mandarin have long intrigued their admirers, the subtle differences between them are equally fascinating. For example, the Mandarin at the ranch, far more often than Wood Ducks, nest near and congregate on shallow upland streams that riffle through boulders and gravel bars on their way to the lakes below. These areas are reminiscent of large parts of the Mandarin's mountainous habitat in the Far East. Perhaps the Mandarin in northern California have found a small available niche somewhere between the turbulent rivers preferred by the native Harlequin Duck and the quiet waters preferred by the Wood Duck.

Also, as if the Mandarin at the ranch were responding to conditions in their colder and more northerly native habitat, they nest three to four critical weeks later than the Wood Duck, a species that evolved in a more temperate zone. During this time, the earlier nesting Wood Ducks claim the best nest sites, and of greater importance, the Wood Duck hens and their broods take over sites that offer the best food and shelter for the ducklings. Many Mandarin broods, arriving later upon the lakes, are thus forced to make the long, dangerous trip down several miles of nearly dry streambed toward the Russian River in search of adequate and less crowded habitat. A number of the hens and their ducklings fail to survive the long journey from the ranch to the river, and the following spring the nests they had used the prior year stand empty. Since Mandarin hens, like Wood Ducks, invariably home to their previously successful nesting sites, an empty nest often means that the hen and possibly her entire brood failed to survive the overland trek.

In other ways the two species seem to respect each other's identity and preserve their own. When they are feeding on the raft on the main lake, one group, invariably the Mandarin, feeds first, just after daybreak, while the other waits along the shore or drifts slowly on the water under the overhanging

OPPOSITE The behavior of Mandarin drakes occasionally appears clownish. A hen looks curiously at their pompous posturing.

willows. After half an hour or so, the drifters gradually approach the raft, and the feeding group slowly departs. As the shift is changing, the raft is often crowded with Wood Ducks and Mandarin together with no evidence of hostility between the species. Within the population of each species, however, competition to protect a mate or ducklings is often intense. In these situations, the behavior of Wood Duck and Mandarin hens is curiously different. A Wood Duck hen drives away other ducks by charging at them with her head lowered and her bill open and threatening. A Mandarin hen responds more cautiously, with beak closed, bobbing and weaving as she strikes at the intruder. When the

The Mandarin drake, unlike the Wood Duck, devotedly guards his nesting hen during the month-long incubation period and is often seen perching nearby as the eggs are hatching.

ducks are on the lakes together, primarily in fall, winter, and early spring, they often rest side by side, Mandarin next to Wood Ducks, on the shoreline, on logs, even perched together in the alders on the lower island. Subtle indications of their individuality can be seen when small groups congregating on logs or in sunny areas along the banks of the lakes drift apart toward nightfall and join discreetly with their own kind.

In spring, a Mandarin drake in the wild can often be found perched atop his nestbox, peering down into the entry hole, at precisely the time the eggs are pipping. Then, unlike the Wood Duck, he stays with his hen and her ducklings during the first week or two after they have left the nest. A Wood Duck drake invariably abandons the pair bond even before the ducklings are hatched and is rarely found with the hen and her brood. The Mandarin drake's attention to his hen and her brood is matched by the Mandarin hen's protection of her nest and its precious clutch of eggs. Whereas a wild Wood Duck hen at the ranch, alert to the presence of predators, often leaves her nest at the slightest hint of danger, the Mandarin hen is less skittish and tends to remain quietly on her eggs even under extreme duress. This attachment of the Mandarin drake to his hen and young brood and the Mandarin

TOP Mandarin hens at the ranch, perhaps less alert to North American predators than the native Wood Duck, set more calmly on their nests, even when outsiders intrude. ABOVE After the ducklings depart the nest, the Mandarin drake remains with the hen and her brood, sometimes until the ducklings have fledged. The Wood Duck drake normally leaves the hen before the eggs have hatched and never tends the ducklings.

hen's protection of her nest, eggs, and ducklings are among the most endearing of the qualities of the species. The Mandarin eggs are more buff colored and slightly larger than the more ivory-colored Wood Duck eggs, and the Mandarin's day-old chicks are larger and heavier than the Wood Duck's. By flight stage young Wood Ducks are equal in size to the Mandarin, and their newly acquired weight may cause them to fly almost two weeks later than the just fledged Mandarin.

The voices and the courting gestures of the two species are also quite different. The sounds of the Wood Duck drake are scarcely audible, while the Mandarin drake is easily identified by his high-pitched staccato bark or, at other times, by his soft, barely audible whistle as he cautiously explores the shoreline. Most unusual, certainly the least often heard, is the grunting of two or more Mandarin drakes together, their heads drawn far back, their chests puffed out, like the posture of the Fantail Pigeon. This is not the "grunt whistle" commonly noted in Mandarin literature, but is a true grunt that normally would be associated in northern California only with a wild pig. Photographers at the ranch, hidden in their blinds, have had to lift the flaps and peek outside to be certain they were not surrounded by the hogs frequently found in this area. It is ironic that such an unbecoming sound comes from such a beautiful bird.

Like most female ducks, the Wood Duck and Mandarin feign injury—often a broken or crippled wing—to lure predators away from their defenseless ducklings. The Wood Duck will only decoy to protect her ducklings. The more protective Mandarin will also decoy predators from a nest full of eggs.

Other very different patterns of behavior by the Mandarin at the ranch may well be the result of their having descended from many generations of birds raised in captivity and long removed from their wild habitat. A striking example is the Mandarin's seeming preference for nesting in man-made nestboxes rather than in natural cavities. We tested this on three remote ponds where Mandarin had nested in nestboxes for years and Wood Ducks never had. One winter the nestboxes were removed and replaced by old snags containing natural cavities. In every succeeding year, Wood Ducks nested in these natural cavities, and the Mandarin never returned, even though an occasional cavity remained unoccupied and available.

Foraging for acorns and other wild foods is another skill the Mandarin may have lost. Under the oak trees beside our cabin, dozens of Wood Ducks probe among the leaves for the fallen fruit—never a Mandarin among them. Mandarin in the Far East migrate north in the spring from the temperate climates of southern China and Japan to the newly thawed wetlands in Manchuria and southeastern Siberia. In the temperate climate of northern

California, as in Great Britain, the Mandarin have exhibited very little, if any, of their natural migrational instinct. In the autumn, the Wood Duck population at the ranch decreases dramatically as they move to the nearby but slightly warmer Central Valley, while the Mandarin population seems relatively unchanged. Perhaps they sense that their unique and exotic appearance, while protecting them at the ranch, would make them a coveted trophy in the lands of the hunter.

The voices of the Mandarin and Wood Duck hens are more similar than those of their drakes. The sharp warning call—the *oeek-oeek*—of the Wood Duck hen is much more audible than the softer warning call of the Mandarin. The coquette call of the Wood Duck hen—*keer-loo-keer-loo*—is unforgettable as it echoes across the water. The Mandarin hen lacks a similar call. When a Wood Duck hen with young ducklings is surprised by an observer, she sits more upright upon the water, makes a series of piercing warning sounds, and raises her crest. In contrast, the Mandarin hen flattens herself on the water, lowering her head and crest as she quietly leads her brood to shelter. The Mandarin hen rarely displays her crest, while the crest of a startled Wood Duck hen is often on full alert.

In courtship the role of both hens is nearly identical: each flattens itself on the water to incite its chosen mate, turning around and around as the drake

TOP Alert to a potential predator, the Wood Duck hen rides high upon the water, crest raised, a signal of alarm to her ducklings.

ABOVE At the sound or sight of danger, the vigilant Mandarin sinks low in the water and silently leads her brood into the shadows of the shoreline.

LEFT Wood Ducks and Mandarin are completely compatible and readily accept each other's presence.

TOP AND ABOVE Mandarin quietly preening, however, may begin to fight among themselves like game-cocks—without apparent provocation, but also without serious consequences.

circles the hen. The gestures of the Wood Duck drake are reserved, whereas those of the Mandarin drake are spirited and, at times, comic. In fact, all the displays of the Mandarin drake—the strutting, huffing, and puffing—are remarkable for their impish character. At other times, from among a group of ducks preening and loafing serenely in the sunshine, two Mandarin drakes start to fight, without apparent provocation, like a couple of gamecocks, jumping a foot or more off the water, striking each other with their wings and claws. Although Wood Duck drakes also skirmish among themselves, we have never seen more than a ritualistic altercation, which does not have the intensity displayed by the seemingly more pugnacious Mandarin.

In the spring, listening to the special sounds and sights of the ducks, we are amazed that, five or six times in less than an hour, a dozen or more Mandarin drakes will fly at high speed down the lake toward the oleander-covered dam. In tight formation, wing tip to wing tip, in almost militant pursuit, they follow a single hen. These are the usually comic Mandarin drakes ardently chasing a mate. This display, like others, is much more aggressive than the more easygoing "courtship flight" of the Wood Duck drakes, also pursuing a single hen but with seemingly far less ardor.

A Mandarin hen "homes" to the same nestbox among the alders that she occupied the previous year.

Throughout the summer, the lakes at the ranch are covered with floating pondweed, and all the ducks—drakes, hens, and fledglings—are drab and lusterless. With the coming of autumn, the first migrating Wood Ducks from the north in their new nuptial plumage spiral in to feed and rest. Then Wood Ducks greatly outnumber Mandarin, but soon, often for a month or more, most of the migrating Wood Ducks and seemingly few of the Mandarin desert our ponds, leaving them eerily quiet. Are they searching the forest floor for acorns or scavenging the unharvested grapes in the acres of vineyards on all sides of us? Or is it that the hunting season is in progress? Although there is very little duck hunting in this area, the Wood Ducks migrating through from farther north undoubtedly have heard and heeded gunshots and hastened to their wintering grounds in the great Central Valley to the southeast, in their passage having been joined by the ranch's resident Wood Ducks and even an adventurous Mandarin. Now, for several months, the roles are reversed, and

Mandarin greatly outnumber the more restless Wood Ducks as they venture to their long-established wintering grounds.

The first major storms of the winter seem to cause many departed ducks to return to the ranch. In the pouring rain, nearly every part of the main lake is covered with Wood Ducks and Mandarin bathing, flapping, and feeding in the weather-swept debris, and nibbling for insects in the swirling currents. Now, the first courtships take place, and at the slightest glint of sunshine on the water, hens flutter to the alders on the island to check for nest sites. These are the sure harbingers of spring, a time when the days will grow longer and warmer, and the fresh foliage of the oaks will brighten the shorelines of the lakes. Once again we will watch for the first ducklings to appear on the dappled pond below the ranch house and will wonder anew about the future of the exotic Mandarin so very far from their native lands.

[L.L.S.]

Nest mates—a golden Mandarin and a silver Woodie—look at the pond below, renewing the hope that the Northern Wood Ducks may continue to live and thrive together on the Pacific Flyway.

THE ABUNDANCE, DECLINE, AND RECOVERY OF THE WOOD DUCK

It was Spring, with dogwoods in bloom. On every hand the swamp encroached on the corduroy road. In that twenty-mile ride, the wood ducks were never out of sight, pair on pair, unafraid, looking up idly as they paddled about. Thousands and thousands of wood ducks were building in that great swampland forest—and nobody then imagined a time when it would not be so.

—CAPTAIN CHARLES ASKINS

OPPOSITE Audubon was as fascinated as other early naturalists by the Wood Duck's roosting and nesting in trees. His vibrant rendering was published in his *Birds of America*. (Courtesy of the California Academy of Sciences, Special Collections)

THIS BEAUTIFUL DESCRIPTION OF THE WOOD DUCK'S abundance was written by Captain Askins during a trip on horseback in 1881 through "the greatest breeding ground of the wood duck—the so-called sunk lands of Missouri and Arkansas." Prior to 1900, most comments regarding the Wood Duck population were, like the one above, extremely encouraging. In 1892, Dr. P. L. Hatch described Wood Ducks in Minnesota in March: "like the rains of the Tropics, they pour in until every pool in the woodlands has been deluged with them." In Wisconsin, a market hunter told Arthur Hawkins of an autumn in 1883: "Wood Ducks were then the most abundant duck with mallards a close second." Writers in southern states such as North Carolina, Kentucky, and Louisiana, just before the turn of the century, were still reporting that the Wood Duck "nests abundantly" and is "common in summer" and "plentiful." Other observers noted that the Wood Duck was an "abundant breeder" in Massachusetts, the "most abundant duck" in Wisconsin, and, in Utah, of all places, a "common resident in the fall."

John James Audubon, in his 1843 *Missouri River Journals,* made frequent and colorful references to the plethora of wildlife in that area. He spoke of "many Wood-ducks" and an "abundance of Geese and Ducks." On one early autumn morning before six o'clock, he recorded that he "ran sixty-one miles, met the steamer 'Satan' badly steered—abundance of Geese and Ducks everywhere."

These were privileged years for observing wildlife in all its beauty and abundance. Not only would the spectacle of thousands and thousands of Wood Ducks be breathtaking in today's world, but so would the sight of birds, as Askins observed, "unafraid, looking up idly as they paddled about." Malcolm Margolin, in his magnificent book, *The Ohlone,* on early Indian tribes of the San Francisco Bay Area, tells of reports in the late 1700s of rabbits so unafraid of humans that they "could sometimes be caught by hand," and of flocks of geese, ducks, and seabirds so enormous they were said to rise "in a dense cloud with a noise like that of a hurricane." Foxes, now excessively shy and secretive, were widespread and common.

At the end of the twentieth century, when most of us have never seen a fox in the wild or a Wood Duck on the wing, it is difficult to comprehend an environment that abounded with mammals and birds within the lifetimes of many of our grandfathers and of all of our great-grandfathers. One hundred to two hundred years is a very short time in which to have converted almost limitless and trusting wildlife into endangered creatures seeking to avoid the very sight and scent of humans.

John C. Phillips, one of the earliest writers to attempt to describe scientifically the status of the Wood Duck and the Mandarin, acknowledged in 1925 that so many authors commented on the former profusion and the then-current decline of the Wood Duck that he felt compelled to mention only a few of the most important accounts. Fortunately, in recent years, several scholars, notably Arthur Hawkins, Henry M. Reeves, and Frank Bellrose, have documented and organized much of these writings. The preceding and following observations of the Wood Duck's abundance and subsequent decline owe much to their efforts.

If wildlife was plentiful in the 1800s, it is almost impossible to imagine

Violante Vanni exaggerated the features of the Wood Duck in his whimsical portrait of the species. His etching, titled *Anatra d'Estate,* is one of the six hundred folios in Xaviero Manetti's five-volume *Ornitologia Metodiche,* published in Florence, Italy, in the late 1700s.

what sights must have greeted the explorers of the North American continent some three hundred years earlier. According to an account written in the early 1500s, "there are many sorts of birds, as Cranes, Swannes, Bustards, wild geese white and gray, and ducks." Concerning New England, Captain George Waymouth wrote of "Fowls—Ducks, Geese, and Swannes—many other fowls in flocks unknown." Whitaker, in Virginia, stated in 1613, "This country besides is replenished with birds of all sorts, which have been the best subsistence of flesh which our men have had since we came. . . . In Winter . . . the rivers and creeks can overspread everywhere with waterfowl of the greatest and least sort, as swans, flocks of geese and brants, duck and mallard, sheldrakes, divers, etc., besides many other kinds of rare and delectable birds whose names and natures I cannot yet recite; but we want the means to take them."

Reports too numerous to mention of the abundance of this period describe wildfowl in general, such as cranes and geese, with which Europeans would have been familiar. Certainly there is no evidence that these seafaring explorers could ever have seen a Wood Duck before, since the earliest imports to Europe were not until 1663 and were invariably kept in small collections on prestigious and presumably inaccessible estates. Furthermore, Wood Ducks were never mentioned in such early writings because at that time they had no generally accepted name. Surely, however, they were prominent among the ducks sighted by visitors to the New World.

The history of the relationship of the early pioneers with what was undoubtedly the North American Wood Duck was assembled by Henry M. Reeves, a waterfowl biologist-historian, and presented in 1988 to the Second

Early explorers and settlers arriving in the New World witnessed what the Native Americans had long seen—impressive flocks of waterfowl, Wood Ducks among them, on the waterways across the continent.

North American Wood Duck Symposium, held in St. Louis, Missouri. With the help of these early explorers and the later colonists, artists, and scholars, one can begin to see the Wood Duck in a brilliant mosaic. In each story, in each report, is information to help paint a single piece.

The earliest known recorded reference to the Wood Duck in North America was made by the Spanish explorer Alvar Núñez Cabeza de Vaca. Although certainly not the first to visit the shores of North America, he is considered the first to record the duck that we now believe was the Wood Duck. Cabeza de Vaca landed on the eastern Gulf Coast in 1527. Eight years and six thousand miles later, he and his small group of four survivors (of the three hundred who started on the expedition) arrived at the Gulf of California—the first explorers to have crossed the North American continent. Cabeza de Vaca subsequently prepared for King Charles V of Spain a detailed account of his remarkable and tragic odyssey, *La Relación*. In describing Apalachen, an Indian village thought to be in northwestern Florida, he noted a wide variety of birds including "geese, ducks, royal drakes, ibises, herons . . . and numerous other fowl." Since the Apalachicola River watershed is currently an area with a high population of breeding Wood Ducks, it is almost certain that Cabeza de Vaca coined the term *royal drake* to describe the crowned, or crested, gaudy Wood Duck drake. Surely he must have encountered thousands of such royal drakes as he crossed the countless wetlands along the Gulf Coast.

Over one hundred years later, in 1675, Christian le Clercq, a Jesuit

Each explorer who encountered the Wood Duck for the first time endowed it with a distinctive name. The *royal drake*, Cabeza de Vaca's descriptive term for the male Wood Duck, was one of many birds he saw when he landed on the Gulf Coast in the early 1500s.

missionary, was sent to work among the Micmac Indians in Gaspesia, in what is now New Brunswick and southeastern Quebec. His *New Relation of Gaspesia* of 1691, written in French, describes the wildlife of the region. The 1910 English translation reads: "The Canadian Ducks are like those which we have in France. There is, however, one different species which we call Canards Branchus [Branch Ducks]: These perch upon trees, and their plumage is very beautiful because of the pleasing diversity of the colours which compose it." Le Clercq may have been the first to write about the Wood Duck other than simply by a name.

Thus, a century and a half after Cabeza de Vaca's royal drake appeared, we have the descriptive name—and, as it turns out for modern taxonomy, a most farsighted name—of "Branch Duck" to identify one of the Wood Duck's unique characteristics. Le Clercq establishes New France (Canada) as the first certain location of the Wood Duck.

Another Frenchman, Dièreveille, arrived at Port Royal, Arcadia (Nova Scotia), on August 20, 1699. Little is known of Dièreveille, not even his first full baptismal name. In 1708 he wrote for his French patron, Michel Bégon:

> The handsomest Birds I saw in that Country are . . . called the Canard Branchus because they roost; nothing could be finer or better combined than the infinite variety of bright colours in their plumage; but this surprised me less than to see them perching on a Fir, a Beech or an Oak & to find that they hatched their young in the hollow of one of these trees where they are cared for until they are strong enough to leave the nest & go with their Fathers and Mothers to seek their food in the water, according to their nature. They are very different from the common Black Duck, which is almost literally that colour without the variegations that ours have; the body of this duck is more slender & they are also more delicate to eat.

As time moves on, so, too, does our knowledge of the Wood Duck and its unique lifestyle. We now have a name, a perching behavior, a nesting style deep in the hollows of trees, and a duck that is "delicate to eat." The pieces are beginning to form the mosaic.

John Lawson, an Englishman, arrived in Charleston, South Carolina, in early September of 1700. It seems that he was a gentleman who had been trained in the sciences and had an urge to travel. Some honor him as the founder of the science of ornithology in North America. With five other Englishmen, three Native American men, a guide, and the guide's Native American wife, they made an extraordinary winter trip of over five hundred miles in only fifty-nine days, arriving in what is now Washington, North

TOP Observing Wood Ducks in the forests of eastern Canada in the late 1600s, missionary Christian le Clercq gave them the fitting name of *Canards Branchus*, or "Branch Ducks." ABOVE By the early 1700s, writings on the Wood Duck noted that they nested in the cavities of old trees.

Carolina, via a route at the base of the Appalachian Mountains. In his journal entry for January 26, 1701, Lawson elaborates on a trip along what is now called the Yadkin River in North Carolina: "On Saturday Morning, we all set out for Sapona, killing, in these Creeks, several Ducks of a strange Kind, having a red Circle about their Eyes, like some Pigeons that I have seen, a Top-knot reaching from the Crown of their Heads, almost to the middle of their Backs, and an abundance of Feathers of pretty Shades and Colours. They prov'd excellent Meat."

In a 1709 detailed listing of birds that Lawson observed, he describes two ducks:

SUMMER DUCK: *We have another Duck that stays with us all the Summer. She had a Great Topping, is pied, and very beautiful. She builds her Nest in a Wood-pecker's Hole, very often sixty to seventy Foot high.*

SCARLET EY'D DUCK: *We kill'd a curious sort of Duck, in the Country of the Esaw-Indians, which were of many beautiful Colours. Their eyes were red, having a red Circle of Flesh for their Eyelids; and were very good to eat.*

BELOW The scarlet eye of the Wood Duck drake, made even more brilliant by the contrasting green crest, attracted the attention of early travelers exploring the wetlands of an unfamiliar continent.

RIGHT Although the female Wood Duck is not as colorful, astute naturalists noted her subtle beauty and her curious nesting behavior.

In retrospect, it seems certain that the duck in the 1701 journal entry and the two "species" listed as the Summer Duck and the Scarlet Ey'd Duck were one and the same. Lawson's entries, like the recordings of those before him, were insufficiently detailed to constitute scientific descriptions. Sadly, depriving us of continuing observations, Lawson was killed by Tuscarora Indians in 1712. But little by little, we have accumulated a quite colorful description of the Wood Duck.

Some two hundred years after Cabeza de Vaca's attempt to name the Wood Duck, Mark Catesby, an Englishman, provided the first illustration, a watercolor engraving, of a Wood Duck drake. There is no doubt that Catesby's Summer Duck is the Wood Duck. Indeed, the watercolor likeness is remarkably faithful considering the state of the art of scientific illustrations of the period. For the first time we can be certain that the subject is the Wood Duck, because we can see it.

Catesby's major opus, *Natural History of Carolina, Florida and the Bahamas* (1731–43), was a true forerunner of scientific writing in America. His illustrations and descriptions reigned supreme until Audubon's work a century later. Most of Catesby's descriptive text, arranged in parallel columns in English and French, focuses on morphology and plumage. He also notes that the ducks breed in Virginia and Carolina, that they make their nests in the holes of tall trees growing in water (particularly cypress trees), and that the adults carry their flightless young to the ground below. This last remark appears to be the origin of the once widely held belief that young Wood Ducks were transported from the nest by the adult, carried either in its bill or on its back. Catesby, usually a reliable writer, fails to give the basis for that statement, which was often repeated by subsequent authors. All contemporary experts believe, on the basis of manifold observations, that the ducklings jump unassisted from the nest, whatever its height, to the water or land below.

Le Page Du Pratz, a French plantation overseer who traveled widely in the lower Mississippi Valley, wrote in 1758: "The Perching Duck or Carolina Summer Duck—their plumage is quite beautiful— and their red eyes appear like flames." This brief notation has great significance today because it is generally accepted as the first use of the word *Carolina* in conjunction with the name Wood Duck. For some mysterious reason *Carolina Duck* was used consistently by early writers in the United States and is currently preferred by almost all writers, particularly aviculturists, outside the country. Thus, for better or worse, Du Pratz left us a lasting legacy.

Dr. Benjamin Smith Barton of Philadelphia, a botanist, wrote of the Wood Duck in his 1799 *Fragments:* "This beautiful species is the gi-gi-tschi-mu-is of the Delaware Indians. It builds its nests in the holes of trees. Attempts have been made to domesticate it; but hitherto, they have not, I believe, been successful."

The Wood Duck appeared as effigies on Native American pottery more often than any other bird except the owl. Abbott noted in 1895: "It is not

TOP AND ABOVE In his two-volume *Natural History*—the first book with colored renderings of American birds—Mark Catesby gives detailed written and pictorial descriptions of the Wood Duck, calling it the Summer Duck, one of the names in common use at the time. (Courtesy of the California Academy of Sciences, Special Collections)

strange that the Wood Duck should have strongly attracted the attention of the Indians. Its wondrous beauty naturally appealed to a savage people, fond of personal decorations and bright colors."

This takes us to the end of the eighteenth century, which we can leave knowing that Native Americans admired the beauty of the Wood Duck, decorated their pipes and bowls with its likeness, and gave it an impressive name.

Generally, with the exception of Barton's writings of 1799, chroniclers of the New World did not begin making scientific contributions to the knowledge of the Wood Duck until the dawn of the nineteenth century.

Thomas Jefferson, the third American president, was keenly interested in wildlife and contributed, at least indirectly, to further knowledge of the Wood Duck. Jefferson's instructions on recording wildlife given to the members of the epic Lewis and Clark expedition before it crossed the continent resulted in the first identification of the Wood Duck in what is now known as the Pacific Flyway. Accounts of five Wood Duck sightings can be found in the expedition's voluminous journals. The most significant entry is probably the one for March 31, 1806: "In the entrance of the Seal River I saw a summer duck or wood duck as they are sometimes called. This is the same with those of the United States country and is the first I have seen since I entered the rocky mountains last summer."

This appears to be among the first explicit uses of the term *Wood Duck*, although the phrase "as they are sometimes called" suggests that it is pre-

TOP The indigenous peoples of North America crafted objects that reveal their familiarity with the Wood Duck. Pottery makers of the Mississippian culture, A.D. 1200–1300, adorned their vessels with the crested head of the Wood Duck drake. (Courtesy of the Dickson Mounds Branch, Illinois State Museum)

ABOVE The Copena culture, active in A.D. 100–600 in present-day Virginia, carved a steatite pipe with a Wood Duck at one end, an owl at the other. (Courtesy of the Smithsonian Institution, Department of Anthropology, catalog no. 211243)

dated by other references to the same name. In fact, the extensive Waldo Lee McAtee papers on common names of birds—nearly seventeen hundred pages of manuscript and ninety file drawers of two-by-five cards—show that the first use of the name *Wood Duck* was by a clergyman, Jeremy Belknap, in his *History of New-Hampshire*, published in 1792.

It is interesting to note that Jefferson's fascination with wildlife was to be followed exactly one hundred years later by that of Theodore Roosevelt, the twenty-sixth president. A revealing insight into Roosevelt's uncanny knowledge of birdcalls, even of British birds, is recorded with admiration and wonder in *The Recreation* by Lord Grey of Fallodon (1918). Lord Grey, coincidentally, was one of the first to raise Wood Ducks and Mandarin together and record the activities of the birds, which were flying in semiconfinement at his small pond in Fallodon, England.

Alexander Wilson was almost assuredly the first to be recognized as a true American ornithologist. He was born in Scotland, the son of a Scottish smuggler. Lacking a formal education, Wilson overcame bewildering odds to become one of America's greatest ornithologists. Like Audubon, he traveled widely and sketched in the field. In his early years in America, he was a teacher and poet. He devoted the remainder of his career to ornithology,

ABOVE In his early study of the Wood Duck, Alexander Wilson groups it with a Long-tailed Duck, a Green-winged Teal, a Canvasback, a Redhead, and a Mallard. (Courtesy of the Smithsonian Institution Libraries, OPPS no. 95-2527)

BELOW As they ended their expedition across the continent, Lewis and Clark saw a duck that they referred to as "a summer duck or wood duck." They were among the first to use the vernacular name of *Wood Duck*.

which culminated in nine volumes titled *The American Ornithology* (1808–17). In volume three Wilson wrote of the Wood Duck:

This most beautiful of all our ducks has probably no superior among its whole tribe for richness and variety of colors. It is called the Wood Duck, from the circumstance of its breeding in hollow trees; and the Summer Duck, from remaining with us chiefly during the summer. Its favorite haunts being the solitary deep and muddy creeks, ponds, and mill dams of the interior, making its nest frequently in old hollow trees that overhang the water. . . . On the eighteenth of May I visited a tree containing the nest of a Summer Duck, on the banks of Tuckahoe River, New Jersey. It was an old grotesque White Oak, whose top had been torn off by a storm. It stood on the declivity of the bank, about twenty yards from the water. In this hollow and broken top, and about six feet down, on the soft decayed wood, lay thirteen eggs, snugly covered with down, doubtless taken from the breast of the bird. On breaking one of them, the young bird was found to be nearly hatched, but dead, as neither of the parents had been observed about the tree during the three or four days preceding; and were conjectured to have been shot.

This tree had been occupied, probably by the same pair for four successive years, in breeding time; the person who gave me the information, and whose house was within twenty or thirty yards of the tree, said that he had seen the female the spring preceding carry down thirteen young, one by one, in less than ten minutes. She caught them in her bill by the wing or back of the neck and landed them safely at the foot of the tree, whence she afterwards led them to the water.

Under this same tree, at the time I visited it, a large sloop lay on the stocks, nearly finished, the deck was not more than twelve feet distant from the nest, yet notwithstanding the presence and noise of the workmen, the ducks would not abandon their old breeding place, but continued to pass out and in as if no person had been near. The male usually perched on an adjoining limb, and kept watch while the female was laying and also often while she was sitting.

This beautiful bird has often been tamed, and soon becomes so familiar as to permit one to stroke its back with the hand. I have seen individuals so tamed in various parts of the Union. Captain Boyer, Collector of the port of Havre-de-Grace, informs me that about forty years ago, a Mr. Nathan Nicols, who lived on the west side of Gunpowder Creek, had a whole yard swarming with Summer Ducks, which he had tamed and completely domesticated, so that they bred and were as familiar as any other tame fowls.

The female has the head slightly crested, crown dark purple, behind the eye a bar of white; chin, and throat for two inches, also white; head and neck dark drab; breast dusky brown, marked with large triangular spots of white; back dark glossy bronze brown, with some gold and greenish reflections. Speculum of the wings nearly the same as in the male, but the fine pencilling of the sides, and the long hair-like tail covers are wanting; the tail is also shorter.

TOP Alexander Wilson was captivated by the dramatic appearance of the Wood Duck drake.

ABOVE In his *American Ornithology*—the first book on American birds published in the United States—he was the first to describe in detail the beauty of the female.

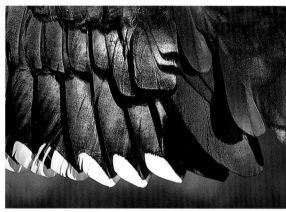

The presence of many Wood Ducks on the ponds and rivers of New England and the reflections of their plumage—even the subtle bronze, blues, and whites of the female (left and above)—inspired Henry David Thoreau to describe the species as a "glowing gem" upon the water.

These descriptions of the Wood Duck paint still other authentic fragments of the ever-enlarging mosaic. In the concluding paragraph above, and in the manner of a true ornithologist, Wilson gives us for the first time a glimpse of the female Wood Duck, although he neglects to suggest that among her waterfowl peers she is considered a beauty. Like his predecessors, Wilson continues to refer to the young leaving the nest with assistance from the hen. It is difficult to explain why many otherwise reliable historical writers report to have seen what, in fact, almost certainly never occurred.

Other pieces of the mosaic appear. Washington Irving, while best remembered for inventing Rip Van Winkle and Ichabod Crane, is also dearly beloved for the more serious writing in his 1835 *Tour on the Prairies,* where he penned a memorable picture of the Wood Duck: "In our course through a forest we passed by a lovely pool, covered with the most magnificent water lilies that we had ever beheld, among which swam several Wood Ducks, one of the most beautiful of waterfowl, remarkable for their gracefulness and brilliance of plumage." In 1847, one of the great American naturalist-philosophers, Henry David Thoreau, wrote: "What an ornament to a river to see that glowing gem floating in contact with its water."

Whereas John James Audubon's paintings are admired worldwide, his prolific and colorful writings are less known. Thousands of wildlife students have described American bird life—but his observations are still some of the

most complete and accurate. His wide-ranging, vigorous mind anticipated by a century the experimental investigation of bird behavior that has played an increasingly prominent role in modern American ornithology. Furthermore, he wrote during a period of the Wood Duck's greatest abundance and at the same time offered clues to the decline that began so soon after his death in 1851.

Although Alexander Wilson was the first true American ornithologist, Audubon, his contemporary, was by far the more famous of the two. Not only were his paintings considered more professional, colorful, and exciting—but so, too, was the man. It is speculated that Audubon, "The American Woodsman," was born April 26, 1785, at Aux Cayes in Haiti on the tropical island of Hispaniola. He adopted North America with vigor and passion and felt nowhere more at home than among the birds of the continent. His rise to fame resulted not only from the brilliance of his art, but from his mastery of the art of salesmanship as well. From the courts and aristocracy of England and France, he cultivated the wealth that would finance his efforts and enrich his own treasure.

Audubon's *Ornithological Biography* was first published in installments from 1831 to 1839. In volume three of the 1835 edition, Audubon wrote what to this day many consider the most poetic and most romantic description of the Wood Duck. It would be impossible to complete the Wood Duck mosaic without quoting some of Audubon's most colorful and revealing insights into the Wood Duck and the time in which he observed it.

TOP AND ABOVE John James Audubon and other ornithologists of his time were fortunate to observe the Wood Duck in the early 1800s, when the species was the most abundant, before unregulated hunting caused its numbers to decline drastically at the turn of the century.

PREVIOUS PAGES In the 1800s, essential nesting habitat was as bountiful as the Wood Duck. Nesting hens could readily find suitable tree cavities not far from water and ample food supplies.

    I have always experienced a peculiar pleasure while endeavouring to study the habits of this most beautiful bird in its favourite places of resort. . . . The mysterious silence is scarcely broken by the hum of myriads of insects. The blood-sucking musquito essays to alight on my hand, and I willingly allow him to draw his fill, that I may observe how dexterously he pierces my skin with his delicate proboscis, and pumps the red fluid into his body, which is quickly filled, when with difficulty he extends his tiny wings and flies off, never to return. . . . On the water, the large bull-frogs are endeavouring to obtain a peep of the sun; suddenly there emerges the head of an otter, with a fish in its jaws, and in an instant my faithful dog plunges after him, but is speedily recalled. At this moment, when my heart is filled with delight, the rustling of wings comes sweeping through the woods, and anon there shoots overhead a flock of Wood Ducks. Once, twice, three times, have they rapidly swept over the stream, and now, having failed to discover any object of alarm, they all alight on its bosom, and sound a note of invitation to others yet distant.

The flight of this species is remarkable for its speed, and the ease and elegance with which it is performed. The Wood Duck passes through the woods and even amongst the branches of trees, with as much facility as the Passenger Pigeon; and while removing from some secluded haunt to its breeding-grounds, at the approach of night, it shoots over the trees like a meteor, scarcely emitting any sound from its wings. . . .

I never knew one of these birds to form a nest on the ground, or on the branches of a tree. They appear at all times to prefer the hollow broken portion of some large branch, the hole of our largest Woodpecker *(Picus principalis)*, or the deserted retreat of the fox-squirrel; and I have frequently been surprised to see them go in and out of a hole of any one of these, when their bodies while on wing seemed to be nearly half as large again as the aperture within which they had deposited their eggs. . . .

The young are carefully led along the shallow and grassy shores, and taught to obtain their food, which at this early period consists of small aquatic insects, flies, musquitoes, and seeds. As they grow up, you now and then see the whole flock run as it were along the surface of the sluggish stream in chase of a dragon-fly, or to pick up a grasshopper or locust that has accidentally dropped upon it. They are excellent divers, and when frightened instantly disappear, disperse below the surface, and make for the nearest shore, on attaining which they run for the woods, squat in any convenient place, and thus elude pursuit. . . . When March has again returned . . . the Wood Duck almost alone remains on the pool, as if to afford us an opportunity of studying the habits of its tribe. Here they are, a whole flock of beautiful birds, the males chasing their rivals, the females coquetting with their chosen beaux. Observe that fine drake! how gracefully he raises his head and curves his neck! As he bows before the object of his love, he raises for a moment his silken crest. His throat is swelled, and from it there issues a guttural sound, which to his beloved is as sweet as the song of the Wood Thrush to its gentle mate. The female, as if not unwilling to manifest the desire to please which she really feels, swims close by his side, now and then caresses him by touching his feathers with her bill, and shews

TOP Beavers build dams that create valuable wetlands.

ABOVE Abandoned Pileated Woodpecker holes are preferred nesting sites for female Woodies.

BELOW A bachelor drake aggressively searches for a mate.

As hunting gained popularity in the late 1800s and early 1900s, finding broods of Wood Ducks and other waterfowl quietly floating upon the lakes and rivers of North America became ever more difficult.

displeasure towards any other of her sex that may come near. Soon the happy pair separate from the rest, repeat every now and then their caresses, and at length, having sealed the conjugal compact, fly off to the woods to search for a large woodpecker's hole.

These lovely excerpts end our historic journey and the piecing together of fascinating, if fragmentary, reports and writings from dozens of pioneer observers into a clear and colorful mosaic of the magnificent Wood Duck—with the sure knowledge of its great abundance and only the gentlest hint of its coming decline.

Hindsight makes it easy to be critical of those believed to have contributed to the decline of the Wood Duck population in the late 1800s and to what many thought was its near extinction by the early 1900s. However, those pioneering days of the eighteenth and nineteenth centuries offered seemingly limitless potential and excitement for the many developments taking place throughout the nation. Wildlife was abundant. Access by steamboats and, later, trains to hunting grounds provided the opportunity for adventurous and fashionable excursions. Although leisure travel for the average citizen had been restricted, new roads that could be traversed by horse and buggy and later by the newly invented horseless carriage began to open previously unknown avenues of recreation. By the late 1800s, sport shooting was popular and highly prestigious, offering camaraderie and pleasure to hunters. The muzzle-loading shotgun was, by this time, in common use. Later, the breech-loading and then the semiautomatic shotgun were invented. The last was the ultimate weapon of destruction for all flying creatures that came within its sights.

Those were days of high adventure, and the possible extinction of any species was beyond comprehension. Today's controversies over the ethics of shooting wild game were almost nonexistent. In the absence of nationwide communications, sportsmen at the turn of the century were each like a hunter in his own isolated blind, taking his share, or more, of birds, not knowing that across the land, thousands, even hundreds of thousands, of hunters were doing the same. They shot and shot until warnings were sounded that soon there might be no Wood Ducks left. It could easily have been a repeat of the demise of the Passenger Pigeon, when the country awakened one day to realize the Passenger Pigeon was lost forever.

Duck hunting in those early days was unregulated. Regulations seem to follow rather than anticipate problems. Wood Ducks were fair game in all seasons—while nesting, raising their young, or migrating. Further, there was no limit to the number of birds that could be taken. During the several weeks when adult Wood Ducks were in full molt and flightless, and when the young were just taking to the air, birds could be harvested by almost any novice with any kind of gun. For the expert marksman, a full bag resulted from only a few rounds of shot. Wildlife was popularly thought to be unlimited, and the killing for "sport" continued.

Market, or commercial, hunting became especially profitable with a number of innovations, including the invention of the repeating shotgun; the use of decoys (both live and wooden), bait (artificially provided grain) to attract birds, and sinkboxes in which the hunter could conceal himself on open water; the availability of improved transportation by which birds could be quickly shipped to cities; and the establishment of specialized game markets. In some locales, huge "punt guns" with bores up to two inches in diameter and weighing up to two hundred pounds were mounted in specially constructed punt boats, which were sculled close to rafted or resting waterfowl. When used in combination with extensive baiting and gunning lights to lure and immobilize the birds, the punt guns could take up to a hundred ducks in one huge blast. Daily bags of up to six hundred birds were common in the mid-1800s. Market hunting was anything but a sport.

Although commercial hunting may have been common, as Henry M. Reeves has noted, the sale of its product was rarely recorded. The most comprehensive description of marketing from the colonial period through the mid-nineteenth century appears in *The Market Book*, with its long subtitle,

Containing the Historical Account of the Public Markets in the Cities of New York, Boston, Philadelphia, and Brooklyn, with a Brief Description of Every Article of Human Food Sold Therein. An account of the West India Company's small store describes "swans, geese, pelicans, ducks, teal, widgeons, brant, coot, divers, and eel-shovellers. These birds were particularly abundant in the spring and fall, and the waters by their movements appear alive with waterfowls; and the people who reside near the water are frequently disturbed in their rest at night by the noise of the waterfowl."

Although the Wood Duck is nowhere mentioned by name, it can be assumed that on the East Coast, where the Wood Duck was so plentiful, it would have been offered for sale. Audubon wrote that in South Carolina "they are bought in the market for 30 or 40 cents a pair. At Boston, where I found them rather abundant during the winter, they bring nearly double that price; but in Ohio or Kentucky, 25 cents are considered an equivalent."

Audubon later wrote that Wood Ducks were readily captured in "figure-of-four" traps, and "that a person in the South traps several hundred in the

When hunting was permitted the year around, flightless adults and fledglings were shot for sport and for sale in urban markets, at prices under one dollar for a pair.

course of a week." In his 1940 report on the birds of Ohio, M. B. Trautman wrote, "Former market hunters and sportsmen found the Wood Duck an abundant transient and common summer resident between 1850 and 1890." The species was said to be so numerous during the summer that a profitable business was made of hunting fledglings and flightless adults in the so-called flapper stage.

Joseph Grinnell, Harold C. Bryant, and Tracy I. Storer provided much information on the Wood Duck in California, where it was once regarded as a common species. In 1918, they wrote that nearly five hundred Wood Ducks were reported sold in the markets in San Francisco and Los Angeles in 1885–86, but during the 1910–11 season, of over one hundred eighty-five thousand ducks sold, only six were Wood Ducks. Clearly the decline of the California population was well under way at the turn of the century.

The Wood Duck, throughout its known history, seems to have been an epicurean's delight. In Holland, ducks were raised so commonly in the seventeenth and eighteenth centuries that they were used by the nobility as table birds. Given the many reports on the Wood Duck's excellence as a table bird, it was doubtless a favorite market duck wherever it was available. However, the relative difficulty of finding and invading the Wood Duck's wooded and often remote habitat made the ducks far less economical for hunters to pursue than, for instance, the diving ducks that concentrated by the hundreds of thousands on the doorsteps of many eastern seaboard cities. The propensity of the Wood Duck to hide in wooded and often inaccessible areas unquestionably protected it from more intensive market hunting over most of its range.

As adventure and excitement aroused the hunting community in those early days, similar enthusiasm motivated farmers, loggers, and miners to cut the forests, channel the streams, and drain the swamps in order to claim the land—often with active backing from the federal, state, and local governments. The land was an even easier target than a bird on the wing. At first, primitive tools, such as the axe, the wedge, and the saw, were used to clear the ancient trees so essential for Wood Duck nesting. Soon, more sophisticated machines— drag lines, steam shovels, scrapers, even dredges—were employed to drain precious swamplands and turn shaded rivers and creeks into barren channels to improve the farmers' harvest. Civilization was on the move, and wildlife was on the run. As a willing and encouraging partner of these seemingly noble efforts at modernization, affluence, and mobility, the government assisted in unleashing the potential of the land in the name of progress. All signals for the human species were positive, but for other species the bells of alarm began to toll ominously. These were the upbeat days of the industrial, agricultural, and recreational revolution—but surely not of the conservation movement.

Although the secretive Wood Duck often went unnoticed in the forest, commercial and recreational hunters were able to locate and harvest the birds, causing their population to plummet by the early 1900s. Regulation of the hunting season and bag limits throughout the first half of the century had a dramatic effect on the Wood Duck's remarkable recovery.

For the Wood Duck, in particular, warnings were being sounded against unregulated hunting (a failure to limit the hunting season or the size of the kill) and heedless habitat destruction. In spite of the slow pace of communications in those days, nearly one hundred years past, the signals of distress were loud and clear.

Reports of the Wood Duck's decline were numerous. Typical warnings that all was not well include the writing in 1901 of naturalist George Bird Grinnell, also a famous hunter: "Being shot at at all seasons of the year, they're becoming very scarce and are likely to be exterminated before long." W. W. Cooke, another highly respected naturalist, wrote in 1906 of the Wood Duck: "So persistent has the duck been pursued, that in some sections it has practically been exterminated." Edward Howe Forbush, a respected ornithologist, wrote in 1925: "Spring shooting, which went on merrily even after the ducks had laid their eggs, brought the species nearly to extinction in the early part of the twentieth century." From Madison, Wisconsin, came the warning that the Wood Duck was "last known breeding, 1890"; from Massachusetts, that the "general decrease began starting in the 70s"; from Buckeye Lake, Ohio, that "a decline was noted starting 1890–1900."

From 1900 to 1924, reports noted a serious depletion of the Wood Duck population. Other reports, making no mention of the decline in numbers of Wood Ducks, were from the more swampy, less populated areas in Florida, Louisiana, and South Carolina, and from the most northerly and more isolated areas including Ontario and British Columbia.

Nonetheless, there can be no doubt that a major decline in the nation's Wood Duck population was in the making and probably gaining momentum with each passing year. Although reports of the Wood Duck's impending extinction were undoubtedly premature, such unintentional exaggerations were instrumental in creating the sense of urgency that led to strong measures to ensure its survival.

The reaction by state governments to the most serious danger—hunting—was both prompt and erratic. The use by hunters of artificial lights to hunt waterfowl was outlawed in several southern states as early as the mid-1800s. Spring shooting was prohibited in New York in 1902. In 1903 two hunters in Minnesota were fined $20,000 each and sentenced to two hundred to three hundred days in jail for attempting to sell commercially more than two thousand birds. The grounds for the suit were "cruel and unusual punishment." However circuitous the legal logic may have been, it was a warning of extreme severity.

Louisiana closed the hunting season for Wood Ducks entirely for five years beginning in 1904. In all, twenty-two states prohibited the hunting of Wood Ducks before the federal government stepped in and negotiated the creative and extremely effective Migratory Bird Treaty with Canada in 1916. Among its other stipulations, this treaty mandated the prohibition of the shooting of any Wood Ducks or eiders for five years. The Wood Duck was the only duck singled out for total protection under the Migrating Bird Treaty Act passed by the United States in 1918. The protection lasted until 1941, when the population increase appeared to justify permitting one Wood Duck in the daily bag in fifteen states. A year later, hunters in all other states were permitted to take one Wood Duck per day during the season.

The recovery of the species in those twenty-two years when the federal treaty prohibited the legal kill of any Wood Duck was nothing short of miraculous. "The comeback of the Wood Duck," Frank Bellrose wrote, "rivals that of the beaver, white-tailed deer, pronghorn antelope, and the wild turkey."

Despite such a magnificent recovery, the Wood Duck population did not consistently follow an upward trend. Consequently, within regulations set forth by the U.S. Fish and Wildlife Service, individual states have, from time to time, lowered the bag limit to suit their needs. This is an essential and ongoing process, and surely a justifiable one for a species that is not truly migratory in the generally accepted sense and is more resident in some states than in others.

When prime wetlands and nest sites are scarce, Wood Ducks pioneer such marginal habitat as small streams and cattle ponds. For three consecutive nesting seasons, a hen raised a brood on this small pond surrounded by stands of acorn-laden oaks.

Such regulations by states and the federal government on the shooting of Wood Ducks clearly have been the major factor in the duck's recovery. By 1963, the legal bag limit was raised from one to two birds. In the middle of the 1980s, the Wood Duck was shot in numbers second only to the Mallard nationwide. Today, nearly one and a quarter million Wood Ducks are taken by hunters out of a total autumn population estimated at seven to eight million birds.

Although regulated hunting ("harvest" is now the euphemism preferred by the U.S. Fish and Wildlife Service) was primarily responsible for the remarkable recovery of the Wood Duck from 1920 to 1942, it was not the only factor. In many ways, the most important element has been the result of the Wood Duck's own natural instincts: In a single nest, it lays a larger clutch of eggs (eight to sixteen) than any other North American game bird. Early in the nesting season, several Wood Duck hens often deposit their eggs in the same nest. It is not unusual to discover a single nest containing thirty or more eggs. Frequently this large clutch of eggs is incubated by a dominant hen to bring forth a brood more than double the size of an average brood. If a Wood Duck's nest is destroyed, it is capable of renesting once, even twice, in a season. It has been more successful in adapting to artificial nesthouses than other tree-hole nesting ducks such as the Common Goldeneye or Hooded Merganser.

Perhaps most important, the Wood Duck can adapt to marginal habitat—or even habitat where the species once prospered but has been killed off entirely. If no hens remain to return to the once-prized areas, young and adventurous hens will begin to explore and nest in these unfamiliar and unoccupied habitats. Much to the Wood Duck's credit, marginal lands often far removed from good breeding areas have also been exploited. Today Wood Ducks can

be found on remote cattle ponds or along the smallest streams in oakland foothills. Food, perhaps more than any other factor, regulates the abundance of a species. Fortunately, Wood Ducks subsist on a broader range of foodstuffs than any other duck except the Mallard. In the Sierra Nevada of California, hundreds of Canada Geese with their goslings can be seen, for the first time, on Lake Tahoe, at an altitude of sixty-two hundred feet. Likewise, Mallards and their broods have for the first time appeared in great numbers on nearby Echo Lake, some seventy-five hundred feet above sea level. In the same way, Wood Duck hens have also pioneered new breeding areas and rediscovered old ones long ago abandoned—thus greatly increasing their nesting potential.

Two special friends of the Wood Duck have cooperated to create additional habitat for nesting and brooding. An explosion in the legally protected Beaver population throughout most of the Wood Duck's range has produced countless beaver ponds, which in time become some of the finest brooding areas for Wood Ducks. Likewise, the Pileated Woodpecker, the largest surviving North American woodpecker, has increased in numbers in approximately the same nesting range as the Wood Duck. Every year they hollow out new nests and abandon old nests that ultimately decay and rot to the size and configuration required by nesting Wood Ducks, providing badly needed nesting opportunities in America's newly maturing forests.

In spite of these adaptations, destruction of wetlands, food resources,

An increase in the Beaver population throughout much of North America has added innumerable shallow, wooded ponds ideal for Wood Ducks to raise their ducklings. Conservation and hunting organizations have worked together to preserve similarly valuable habitat for Wood Ducks and other waterfowl species.

The Wood Duck, which was believed to be dangerously close to extinction at the beginning of the twentieth century, has made a remarkable comeback. The autumn population is currently estimated at between seven and eight million birds.

and nesting facilities for the Wood Duck is the ultimate threat to its existence. Habitat deterioration is a slow and almost invisible process, but it is the drop of water that ultimately consumes the stone. Although degradation of habitat has been under way for over one hundred years, it has not yet reached the critical stage for the Wood Duck's survival. A multitude of programs, starting slowly in the late 1930s, have ameliorated habitat destruction for all waterfowl, so the future is encouraging—at least for the Wood Duck.

Whereas the success of hunting regulations was well publicized, the effort of one remarkable man is less known. Shortly after passage of the Migratory

Bird Treaty Act in 1918, Alain C. White became interested in the plight of the Wood Duck. Trained as a botanist, he decided upon a unique course of action to assist the endangered Wood Duck. In 1922 he imported captive Wood Ducks from Belgium and brought over from England a professional gamekeeper to supervise a program of raising the ducks on his beautiful Bantam Lake in Litchfield, Connecticut. Up to four hundred ducklings were banded and released each year. By the time the program was concluded in 1939, a total of nearly three thousand birds had been released from the Bantam Lake breeding area, and bands from these birds had been returned from fifteen states and Canada. S. Dillon Ripley, past secretary of the Smithsonian Institution, refers to White, his old-time neighbor, as the father of Connecticut conservation and concludes that these Wood Duck releases "probably had a seminal effect on the restoration of the species in the wild." One man had certainly made a difference.

In the fifty years following the initial comeback of the Wood Duck, much more has been done to ensure its survival. Of major importance has been the creation of scores of private organizations whose actions directly or indirectly benefit the Wood Duck and wildlife in general.

The sale by the federal government, and later by state governments, of duck stamps that must be purchased by hunters has funded state and local conservation programs and, in the process, has provided the nation with

In the last half of the twentieth century, the use of man-made nest-houses to replace disappearing natural cavities has helped add several hundred thousand Wood Ducks to the populations on all of the North American flyways.

beautiful wildfowl art. In what seems almost a contradiction in terms, duck hunters are being recognized as one of the great benefactors of wildfowl. Thousands of acres of private wetlands have been financed and maintained by many organizations such as Ducks Unlimited and by hunters in duck clubs nationwide, which also provide much-needed natural food for ducks in migration. Private duck clubs, in general, play an important role in enforcing hunting regulations among their members, often setting their own limits below those allowed by law. It is often said that if the hunting of ducks in North America were to be too severely limited or forbidden altogether, such wetlands would be taken over and drained or filled by farmers and developers, thereby sounding the death knell for many species of wildfowl on the continent.

As a nonhunter, I have always delighted in the thoughts of Aldo Leopold, one of this country's early great naturalists and conservationists and also an ardent hunter:

> Hunting for wild game is a sport for nature lovers—those who sit in awe before the splendor of the rising sun, who enjoy the animals, the birds and all wildlife about them. Hunting is not primarily for killing. Rather, it is for maintaining the proper balance of nature in conformity with all the laws that regulate this process. Like a good fisherman take only what you need and release the rest, in this case releasing the rest means letting them pass by unharmed. I hope this sport brings you as many hours of pleasure as it has to me.

Whereas most efforts to protect waterfowl have been of general benefit to all species, one in particular has been designed almost exclusively to benefit the Wood Duck: the unique artificial nestbox or nesthouse. Wood Ducks are fortunate to be one of the few ducks to adapt easily to such man-made nestboxes. Yet, according to Frank Bellrose, one of the pioneers in nestbox design and deployment, "Up until the early 1950s, nesthouses played an insignificant role in the comeback of the Wood Duck; there were just too few nesthouses to be meaningful." However, he estimates that today the number of nesthouses, constructed and installed by hundreds of federal, state, and local agencies and by thousands of private groups and individuals, may total over one hundred thousand. They alone may contribute annually one hundred fifty thousand additional ducks at flight stage, important even to a total fall population estimated at seven to eight million Wood Ducks. Nesthouses, properly designed, located, and maintained, are potent insurance against the possible depletion of natural nesting sites. Equally important is the satisfaction and joy they bring to the many individuals who labor happily each season to provide a safe, clean, and inviting nest for the ever-seeking hen.

The comeback of the Wood Duck, aided by many elements in society, has been nothing short of phenomenal—and while the populations of other important game ducks such as the Mallard, the American Black Duck, and the Northern Pintail are declining, the number of Wood Ducks continues to hold its own or even to register a modest increase. Assuming that nature does not wreak catastrophic damage on waterfowl, that federal and state conservation agencies continue their program of securing and improving precious wetlands, that the length of the hunting season and the hunter's harvest are prudently regulated, and that other habitat and nesthouse programs continue to grow, the future for the once-threatened Woodie looks very bright indeed.

[L.L.S.]

Centuries after the early explorers first saw Wood Ducks in the New World, this elegant species is once again abundant in North America.

CHAPTER THREE

THE HOME AND HABITS OF THE WOOD DUCK

Ten thousand years ago on the North American continent, retreating ice sheets and glaciers sculptured a landscape that, over time, created a paradise for wildlife like none other in the world. Unlike other continents with huge inland areas of desert, tundra, or arid, woodless plains, most of the North American continent more closely resembled a seemingly boundless wildlife reserve.

As recently as the eighteenth century, reports describe the Atlantic coastline as being replete with shorebirds and waterfowl, and the eastern deciduous forests as teeming with wildlife. On the prairies to the west, all manner of waterfowl shared a rich environment dotted with lakes, rivers, and potholes scoured from the earth by the long-departed glaciers. Vast herds of Bison and Pronghorn grazed the plains. Farther west, giant mountain ranges harbored Grizzly Bear, Mountain Lions, Mountain Goats, deer, and elk. The Pacific shoreline, its estuaries, and its fertile river valleys hosted a world of wildlife: waterfowl, seabirds, seals, sea lions, and Sea Otters, California Condors, and Golden Eagles.

Across the entire continent were vast areas of wilderness and networks of rivers and streams, lakes and swamplands, interspersed throughout woodlands and meadows, offering magnificent habitat for the forty-three species of

What a beautiful creature is this Beau Brummel among birds and what an exquisite touch of color he adds to the scene among the water hyacinths of Florida or among the pond lilies of New England.

—Arthur Cleveland Bent (1923)

OPPOSITE *Regal Duck, King Duck, Bride Duck,* and *Rainbow Duck* are among the names used since the mid-1500s, when Cabeza de Vaca is believed to have described the Wood Duck as the *royal drake.* The many names given to the species over the centuries have attempted to capture its beauty and its bearing.

83

Seeing the Wood Duck among the water hyacinths of Florida, Arthur Cleveland Bent, one of the early admirers of the species, described it as the "Beau Brummel among birds."

waterfowl that are native breeders in North America. Of the approximately one hundred forty-five species of ducks, geese, and swans in the world, no duck is more exclusively a resident of the forty-eight contiguous states than the Wood Duck. It nests naturally in every state and in significant numbers in more than forty and makes important forays to nest in the spring just north of the Canadian border and, to a lesser degree, flies south to winter in Mexico. Although the continent is home to all of the world's wild Wood Ducks, so secret are their ways and often so impenetrable their habitat that it is impossible to estimate accurately their total population—believed to number, late in autumn, seven to eight million adult and juvenile birds.

Since the early 1700s, scientists and naturalists have studied and written of the beauty and unique characteristics of this North American duck. The striking colors of the Wood Duck drake and the subtle beauty of the hen have long attracted attention. Also, since the Wood Duck is found on as wide a variety of wetlands as any other species of waterfowl—nesting on lakes, rivers, creeks, and ponds, and on every other wetland that supports nest sites—it is often close to human habitation, where it has adapted to live most frequently in man-made nesthouses. Occasionally, it even tries to nest in barns or in residential chimneys. In paintings, sculptures, photographs, and literature, artists and writers have expressed their affection and admiration for the Wood Duck—a bird so arrestingly beautiful, but also so secretive and shy, that relatively few people are fortunate enough to see one in the wild. While the mystique and elegance of the Wood Duck have captured the curiosity and affection of nature lovers everywhere, this attention has not always worked to benefit the Wood Duck. Hunters have found the Wood Duck a tasty table bird, and taxidermists have profited from its beauty. Even today flytyers cherish the exquisitely barred flank feathers from which they fashion a number of fly patterns.

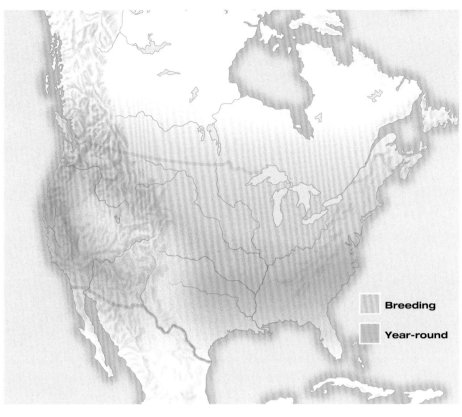

Breeding

Year-round

Range of the Wood Duck in North America.

The migration of the Wood Duck is unique. Of all the North American waterfowl, the Wood Duck is the only species with a portion of its population (approximately two-thirds) that is essentially migratory and a significant southerly portion (approximately one-third) that is mostly nonmigratory, or resident. Wood Ducks inhabit three principal flyways—the Atlantic and Mississippi Flyways, separated by the twelve-hundred-mile-long Appalachian Mountains, and, far to the west, the Pacific Flyway. The two eastern flyways join in a giant horseshoe at the southern tip of the Appalachians in Alabama. By contrast, the Pacific Flyway, which accounts for less than two percent of the Wood Ducks on the continent, is separated from the rest of the country by the massive Sierra Nevada and Rocky Mountain ranges. So nearly complete and long-standing was the isolation of the Pacific Flyway that it was often speculated that the Pacific Wood Ducks might have developed the characteristics of a subspecies.

However, between the Mississippi and the Pacific Flyways, in the plains states, lies the Central Flyway, encompassing in more recent times a minor but slowly growing population of Wood Ducks. From this increasing population on the Central Flyway, Wood Ducks already have found their way to the west, bringing new and possibly welcome bloodlines to the Pacific population.

Since the secretive and shy Wood Ducks seek refuge in remote and often

impenetrable forest wetlands, it is a rare and memorable experience to hear or see a flock of them at dusk or in the early dawn, turning and twisting against the sky, the wind gusting through their braking wings as they splash onto a peaceful pond, bringing it alive with the unmistakable sounds of the Wood Duck hen. So distinctive are the sharpness and shrillness of her call of alarm that it long ago earned the Wood Duck the colorful name of *squealer*, which in early times was often mistakenly attributed to the drake, whose calls are actually far less audible.

A bird of quiet ponds, deep woods, and dark forests, the Wood Duck inhabits territory whose glossary of names is as beautiful as the duck itself. Rivers and streams throughout the flyways are referred to as *riverine* and their shorelines as *riparian*. Unique to many of the southern states are inland swamps and marshes often described as *palustrine*. Lakes housing Wood Ducks throughout the land are *lacustrine*. Oxbows, Baldcypress and tupelo swamps, overflow bottomlands, alluvial flood plains, greentree reservoirs, and beaver ponds are among the many colorful designations for the almost limitless variety of freshwater wetlands inhabited by the Wood Duck.

OPPOSITE A female Wood Duck rises spectacularly from the water. ABOVE The drake follows her in an almost vertical ascent. In proportion to its size, the Wood Duck has the largest wing of any species of North American game duck.

Long ago the sheltering environment of swamps and marshes of all kinds, beaver ponds, and overflow bottomlands were probably the Wood Duck's primary habitat. However, over time these vast lowland areas have been deforested and drained for farmland, until today Wood Ducks are found in greatest numbers in the nearly one million miles of rivers and streams of North America.

An essential characteristic of all Wood Duck habitat is fresh water, mostly slow moving, and preferably shallow, and well sheltered by trees and shrubs, yet bathed in sufficient sunlight to create an abundant food supply to satisfy the energy needs of adult ducks and their ducklings. Equally important are the ancient trees for nesting and the cover needed for Wood Ducks to hide from countless predators—and to provide them with sun-dotted areas for the preening and loafing that offer rest to an otherwise restless lifestyle.

As the seasons change, Wood Ducks move north or south on their flyways, leaving old habitat for new, and altering their diet and even their plumage. Watching them throughout the seasons is to observe them as through a giant kaleidoscope—where they are part of a constantly changing collage of nature's patterns.

LEFT Preening is an essential part of courtship.

BELOW The drake tips for acorns while his mate watches for danger.

OPPOSITE, FROM TOP A wide variety of nuts, seeds, and berries make up the Wood Duck's diet: acorns, bur reed, Wild Rice, sedge, Black Tupelo, and Flowering Dogwood.

FALL For Wood Ducks everywhere, late summer and early fall are the seasons for renewal. The summer molt has mostly passed. By early fall the drakes are flying again and soon the hens will regain their flight feathers and join them in the sky. Shortly adults of both sexes, and only slightly later the juveniles, will assume their splendid nuptial plumage. By late August or early September, almost three million adults plus more than four million juveniles are adapting to the often bright and colorful days of autumn when newly feathered birds engage in the intricate courtship ritual that results in the early pairing of some adults before winter. This prenuptial ritual consists of as many as a dozen or more colorful displays by the drake and somewhat fewer by the female—often followed by the hen lying low in the water, head and neck outstretched, inciting copulation. In contrast, post-pairing display primarily involves affectionate preening, by the hen or the drake or both, of feathers around the bill and eyes. After pairing, the two ducks often swim slowly away with the hen leading. Hours are spent on the shoreline or on shallow bars searching for the acorns so abundant at this time of the year.

Wood Ducks are omnivorous, consuming as wide a variety of food as any other duck. Across the continent acorns are the preferred food in an adult Wood Duck's diet. Wood Ducks can often be seen in shallow wetland waters, heads down, tails up, tipping or dabbling for acorns dropped from overhanging branches or scouring the forest floor in search of freshly fallen fruit. More than a dozen other nuts, seeds, and berries are savored by Wood Ducks. These include hickory and beech nuts, Black Tupelo, Wild Rice, duck-weed, sedge, bur reed, dogwood, and numerous other plants and tubers. For long periods, while not feeding, the birds stand on one leg on fallen trees or stumps, preening or resting, often half asleep with their heads tucked behind their wings—but always alert to any danger.

Farther north, as the weather turns colder, the time for migration to warmer southern habitat approaches. Birds move restlessly about. Many more birds are on the ponds, and small flocks continually leave as others arrive. During this premigrational movement in and about the nesting areas, young and adult Wood Ducks pass daily up and down waterways and expand their areas of flight. Many ducks return periodically to their natal grounds, and most flights are normally not more than a few dozen miles. The number of ducks in a given area remains fairly constant, but the makeup of the group varies greatly as new birds enter the area and others depart.

By early September, in the far north the first ponds and streams are

ABOVE Each autumn, wildlife watchers witness the glorious sight of flocks of ducks flying south for the winter.

BELOW Wood Ducks are often hunted on rivers and streams during fall migration.

beginning to freeze. The time for serious migration has arrived. In the Atlantic Flyway, the first Wood Ducks to start their journey south are those from as far north as central Quebec and Nova Scotia, which begin to leave in early September and continue to depart until mid-November. Birds from Maine follow shortly and are soon joined by other birds from New England, from the slow-flowing rivers and millponds of Connecticut and Massachusetts. By December, large numbers of migrating birds are leaving or passing through the mid-Atlantic states, from New York to the Carolinas.

In the Mississippi Flyway, the first migrating Wood Ducks leave Ontario in southern Canada by mid-September, followed by birds from Minnesota, which is home to more breeding Wood Ducks than any of the forty-eight contiguous states. By October birds begin to leave Indiana, where the St. Joseph River and the Muscatatuck River harbor among the most dense populations in the state. Joining the exodus are birds from Kentucky's streams, which may entertain more breeding Wood Ducks per mile than streams found elsewhere in the country. From Illinois, home to over two hundred thousand Wood Ducks, birds wing southward from waterways such as Quiver Creek and the slough of Nauvoo.

In the somewhat greater warmth of the Pacific Flyway, early winter

migration is an even more leisurely affair. Probably fewer than half of the Pacific Wood Ducks are migratory, and these ducks from western Canada, Washington, and Oregon head mostly for the large Central Valley of California, a few flying as far south as the Mexican border.

The southerly migration in the fall coincides largely with the opening of hunting seasons in the various states, the reason being, of course, that the best hunting potential is when birds are flying overhead in migration. Birds from the north heading south are the first to feel the effects of the hunters' guns. The opening days of duck season bring forth the greatest turnout of hunters, and a Wood Duck from Minnesota may endure as many as seven opening days and partial seasons as it passes from state to state and may suffer, in total, three continuous months of shooting. By contrast, a Wood Duck resident in Louisiana encounters hunters on only one opening day and is shot legally for a maximum of thirty to forty-five days, depending on the length of the hunting season that year.

From coast to coast those migrating birds fortunate enough to have survived the long and arduous flight through storms and shot arrive at their age-old wintering grounds in the south.

WINTER In the east, the Wood Duck's primary wintering grounds are the six states of the Deep South—from Louisiana in the west through Georgia and North Carolina and South Carolina to the east, and including the superb wetlands in the upper half of Florida. Of slightly less importance for wintering waterfowl are the southern parts of Virginia, Tennessee, and Arkansas, and the wetlands of eastern Texas. Into these areas from September into January, Wood Ducks arrive in an unbroken procession of small flocks. As winter progresses the Wood Duck population in the Deep South increases dramatically as several million birds from the north having survived the often perilous flight join well over one million year-round resident Wood Ducks.

The southern wetlands in winter are a Wood Duck paradise. Sluggish rivers and streams, alluvial hardwood bottomlands, and deep swamps of Baldcypress, tupelo, and gum trees offer forested wetlands rich with acorn, beech, and pecan mast and other nutrients. In winter, the leafless branches of deciduous trees offer little protection in areas of open water. In this season the ducks are extremely wary—whether of hunters or predators—and seek more densely covered areas. Wetland vegetation providing shelter for the ducks includes the ever-present willows, bulrushes, Buttonbush, swamp privet, arum, arrowheads, lizard's-tail, spatterdock, water-primrose, and Water Smartweed. These are the days of the Wood Ducks' Mardi Gras, a festival allowing those ducks arriving from the north to escape the rigors of winter and providing for both visitors and residents the glorious days of courtship and, finally, the taking of a mate. A number of adult birds have already paired in autumn on their breeding grounds, prior to migration.

Of the recently paired adults, however, slightly less than one-fifth have fallen during migration to the guns of hunters. With so many ducks concentrated in these southland swamps and ponds, predators inhabiting the same waterways—Raccoon and Mink among others—make heavy inroads on the wintering birds. Winter storms and disease likewise take their inexorable toll.

In total, almost one-half of all adult and yearling Wood Ducks will have fallen to hunters' guns and natural disasters. Newly arrived yearlings, never paired before, are in a courting mood. Winter, when all the birds are in their finest nuptial plumage, is a time for courtship. Older birds seek to replace lost mates, often of many seasons past, and yearlings enjoy the excite-

BELOW, FROM TOP Deep swamps of Baldcypress and Black Tupelo are ideal wintering grounds. By early November, ponds in the far north begin to freeze. Even in southern states, lakes ice over in bitter winter storms.

ment of their first flirtations while learning much about courtship from the more experienced adults.

While the birds seek rest, recuperation, and romance in the wet and wooded confinement of these southern states, there is continuous movement within the area. Birds shift constantly to search for food, to seek open waters as colder areas freeze, or simply to escape the violence of the winter winds. With all these comings and goings, birds from Minnesota may pair with others from as far away as Maine, or birds from Alabama with those from Iowa or Texas, thus ensuring the yearly outcrossing that renews the vitality of the species. Despite such gatherings of Wood Ducks originating in more than thirty-five states and wintering in an area only one-fourth that size, and even with millions of spectacular drakes in their eye-catching nuptial plumage, few, if any, Wood Ducks are to be seen as they congregate on their backwater hideaways. Yet in their own way, they are enjoying the leisure and vicissitudes of this special season of the year. Only the birds isolated far to the west on the Pacific Flyway celebrate a somewhat calmer and warmer winter holiday.

SPRING By early March, the pairings of most Wood Ducks will have occurred. Then, as many as two-thirds of all the wintering pairs begin heading north, some as far as the New England states on the Atlantic Flyway and Minnesota on the Mississippi Flyway. More than two hundred thousand Wood Ducks cross the border into southern Canada, including Ontario, Quebec, and New Brunswick. In the Pacific Flyway, the spring destinations of the more modest numbers of migrating Wood Ducks are Oregon, Washington, and British Columbia.

It is a phenomenon of nature that the direction, distance, and final destination of each pair in their spring migration is seemingly preordained. For many the flight to the spring nesting grounds is over familiar terrain, as older

TOP A nesting Wood Duck hen often has great difficulty finding a suitable tree cavity to nest in. ABOVE A prairie-nesting duck, such as the Mallard, needs to locate only a flat patch on the ground.

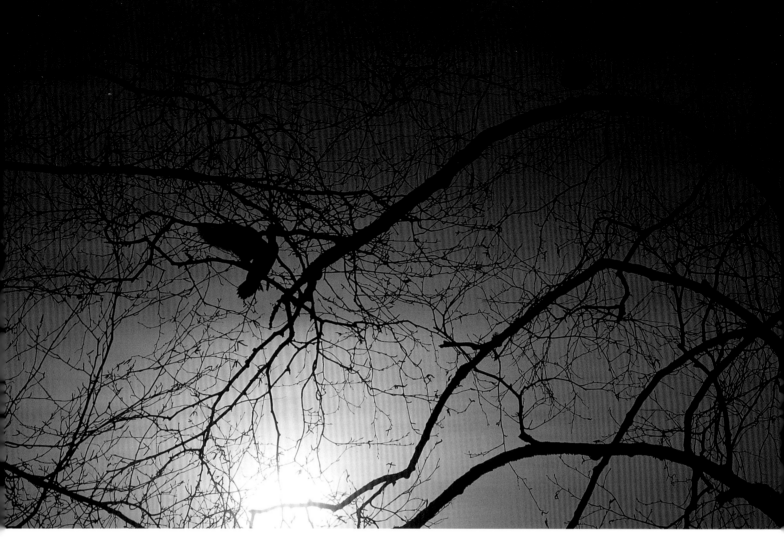

ABOVE The search for nest sites begins at sunrise as pairs of ducks fly high into the treetops.

BELOW Many species compete for the nest sites used by Wood Ducks. Here, a Western Screech Owl using a Wood Duck nestbox has brought back mice for her owlets.

hens, with a keen "place memory," seek to return to last year's nest sites, or even last year's nest, if they were successful in delivering at least one live duckling into the wild. A large number of yearling hens likewise seek to nest close to the places where they were hatched. This homing ability of female Wood Ducks is known as philopatry, and although still not entirely understood, fortunately for the Wood Duck, it is more highly developed than in most prairie-nesting ducks and is of great importance to the survival of the species.

A prairie-nesting duck need find only a small piece of land near water, a source of food, and a bit of shelter to construct its nest. However, a Wood Duck hen nests only in special cavities inside a tree—or in a man-made nestbox. Natural cavities are often scarce, far from water, and difficult to discover. The ability of a Wood Duck hen to home to an area or even to the very nest she used successfully the previous year saves countless, possibly fruitless hours of prospecting in new areas for such an unusual nesting site. Likewise, female yearlings following their hens, or homing on their own to their natal nest sites, are likely to find suitable tree cavities or nestboxes in the same area.

It is speculated that aeons ago the Wood Duck, like other ducks in prehistoric times, was a ground-nesting duck. Over many thousands of years of evolution, the Wood Duck may have moved its nest from the ground high into the trees, as it sought to escape the concentrated pressures of predators and the destruction of nests and food sources by the ever-changing water levels of the swamps that were its favored habitat.

In the process, the anatomy of the Wood Duck may have gradually changed. The shape of its feet and the placement of its legs, more forward on its body than on prairie-nesting ducks, give the Wood Duck great agility to perch and move about the branches of trees. Then it was only natural that in some ancient oak, Baldcypress, or tupelo, it would find a cavity in which to nest with greater safety. Even the young ducklings evolved a set of sharp, kittenlike claws to climb out of the nest cavity, often two to six feet in depth. To navigate swiftly in the half dark of dawn through dense forests, the Wood Duck, along with its cousin the Mandarin, developed one of the largest eyes in relation to its body weight of any other waterfowl. Likewise, to maneuver through tangled branches, where other ducks rarely fly, Wood Ducks and Mandarin developed a relatively broader wing and longer tail than most other ducks.

Courtship and mating take place on and even under the water.

Having adapted to wooded wetlands, the Wood Duck faced only limited competition from other species of ducks for natural cavities in which to nest. Fortunately, its potential competitors for nests, Hooded and Common Mergansers, Common and Barrow's Goldeneyes, and Bufflehead, seek different habitat and are therefore only marginally found nesting in Wood Duck areas. Although other ducks normally do not intrude on the Wood Duck's breeding grounds, many other species of wildlife, including squirrels, small owls, flickers, woodpeckers, starlings, and several species of bees, covet and monopolize the same type of nest. So locating and securing a safe and private nest cavity is a serious undertaking for a Wood Duck.

In spring, the female Wood Duck is the dominant member of the pair. The Wood Duck drake, as with most drakes in the duck family, follows his hen as she seeks a new nest site in the forest. Thus, a drake paired last spring with a hen nesting in northern Minnesota may find himself this season following a mate to the bottomlands of Florida—one thousand miles from his last year's nest. Whereas the hen's destination in spring may be predetermined by her propensity to home to last year's nest site, the drake's is as uncertain as that of any early explorer of this vast continent.

Other pairs of Wood Ducks, led by resident females, remain to nest again in the South on the rivers and swamps of the Carolinas, and on the swamps and beaver ponds of Georgia, Alabama, and other states as far south as Florida and west into central Texas. Weather primarily determines the time of nesting. In the southern states, nearly a million Wood Ducks begin their nest search two to three months earlier than birds traveling to the colder climates of the north. First eggs have been noted in southern states as early as January, whereas first laying in northern states commonly takes place in early April.

Regardless of the geographic area to which the Wood Ducks return in spring, the processes of searching and finding a nest are almost identical, even though choice of habitat varies and some nests are established earlier

BELOW A nestbox is opened to check the setting Wood Duck hen. RIGHT The hen is easily lifted from her eggs and banded.

than others. For the nearly three million nesting birds, two substantially different nest styles are available. In areas with old-growth hardwood forests, natural cavities may provide the unique style of nest that Wood Ducks have long enjoyed. To be attractive to Wood Duck hens, a nesting cavity must have certain essential characteristics. The most important are visibility to searching ducks; height above the ground or water ranging from five to fifty feet; adequate interior size and depth to accommodate a nesting hen; clear and easy access to an entry hole of acceptable size; sufficient rotted wood, leaves, or litter with which to cover the first five to seven eggs; a distance not much greater than a mile or so from the nearest stream, pond, or other source of food and shelter for the ducklings.

In the north, such nesting trees include maple, aspen, and basswood. In the central latitudes, there are various oaks, beech, willow, Boxelder, sycamore, and elm, while in the south, there are Water Tupelo, willow, cypress, and some thirteen varieties of oak that are able to grow with their roots submersed in water for differing lengths of time. Surprisingly, most of the natural cavities used by Wood Ducks are in live trees—not in dead snags, as one might expect.

Many areas where natural cavities were once abundant in the nineteenth

Ornithologists and wildlife artists have described the remarkable similarities of the female Mandarin and Wood Duck. Both male and female are the only ducks whose primary, or flight, feathers display a silver-white coloration on their leading edges.

A hen searching for a nest site selects an old nestbox in which to hatch her ducklings.

and early twentieth centuries have been heavily logged, giving rise to the belief that a shortage of nest sites was, in part, the cause of the apparently severe decline in the Wood Duck population in the late nineteenth century. The decrease in natural sites was more obvious near major centers of human population, mostly in the northeast. Despite these localized shortages, it is believed that approximately ninety-five percent of all Wood Duck nests are still found in natural cavities—and in recent decades, for many reasons, the Wood Duck population continues to prosper.

The second type of nesting accommodation attractive to Wood Ducks is the man-made nestbox, or nesthouse, designed to simulate an ideal natural cavity and to provide nesting opportunities where natural cavities are scarce (see Appendix I). Although just over one hundred thousand man-made structures are estimated to be in use today, they serve a purpose out of proportion to their numbers because they have been the chief source of abundant and carefully documented knowledge about the Wood Duck— knowledge that would not be available through studies of natural cavities, which are extremely difficult to locate and monitor.

Practically all the literature about Wood Duck nest sites pertains to man-made structures. Few studies have evaluated the location of tree-cavity nest sites and the characteristics of the Wood Duck's use of these cavities. Because of the ease of access to nesthouses, nesting hens are quite easily caught on the nest and tagged or banded. Their movements are subsequently recorded, and their eggs weighed, counted, and otherwise studied. Day-old ducklings, without being harmed, may be tagged in the webs of their feet, and their movements, even their homing capabilities and their survival rate, can be evaluated. In short, much of what is known today about wild Wood Ducks is the result of careful and humane studies originating in man-made nesthouses, including the duck's nesting habits, quantity and fertility of eggs laid, period of incubation, how ducklings leave the nest, number of successful nests, number of nests destroyed or deserted, migratory patterns, knowledge of predators, and, finally, even population estimates. Gently handled, incubating hens may often be removed from the nest and their eggs candled and marked, sometimes daily, without causing the hens to desert.

It is one of nature's ironies that such a shy and reclusive duck, which naturally nests in a well-hidden cavity deep in the forest, would be so easily attracted to an artificial cavity installed often in the heart of a town or on the outskirts of a large city. Nestboxes are even located successfully along freeways—sometimes only a few yards from, or even on the side of, an occupied dwelling. For some unexplained reason, the normally shy and retiring Woodie seems content to nest in close contact with human habitation. Perhaps the nesting hens recognize, intuitively, that a well-designed and judiciously installed nestbox close to human activity provides greater protection against predators than does a natural cavity. This apparent trust has endeared the Wood Duck to all who experience it.

Whether a Wood Duck is inspecting a natural cavity or a nestbox, the routine is identical. The hen leads the search, flying from tree to tree until, at last, she finds an inviting nesthole. She clings to the surface beneath the hole, much as a woodpecker does, while peering into the entry, sometimes for a minute or more before entering it. If for any reason that nest is not suitable, she leaves the site to prospect elsewhere. Or she may enter the cavity or nestbox, remaining there for a moment, or for as long as several minutes, often scooping out a depression in the litter. Occasionally, the drake, usually in close attendance during the search, flies to another tree as if inviting his mate to inspect a cavity she might have overlooked. The search may continue for only a day, or for a week or even a month, with the final decision of a particular nest apparently made by the hen alone. Rarely does a drake ever enter a nesthole.

Once the site has been selected, the pair often flies to the nest

During the search for a nest site, the drake is in close attendance, but rarely enters a nesthole.

TOP When a hen leaves the nest to feed, she blankets the eggs with her down. Uncovered eggs indicate that the hen was frightened off the nest by a predator.

CENTER A dump nest often has thirty or more clean, shiny eggs laid by two or more hens. The bits of down here suggest that the eggs will be incubated by an adopting hen.

BOTTOM A drop nest contains soiled eggs left in disarray. They will never be incubated.

immediately after sunrise of the following day. As the drake stands guard in a tree close-by, the hen enters the nest to remain for a few minutes, up to a half hour or more, as she lays her first egg. Upon leaving, she covers it with rotted wood litter or shavings, as she does each egg laid on succeeding days until the sixth or seventh egg. At that time or shortly before, she begins to pull or shed the downy feathers from her breast. At first she mixes her down with litter to cover her eggs when she is off the nest. By the time the last egg is laid, she has accumulated a magnificent blanket of down often more than an inch deep, with which she covers her eggs to keep them warm whenever she voluntarily leaves the nest. If frightened off the nest, however, she leaves her eggs uncovered and often in disarray and soiled. The resulting scent is thought to repel invading predators.

A normal clutch for a single Wood Duck hen is from ten to sixteen eggs, with approximately twelve eggs being the accepted average. More than sixteen eggs in a nest is presumed to be what is called a "dump nest," the result of two or more hens of the same species laying in the same nest. Over thirty eggs are often found in a single nest, and one report tells of a dump nest with fifty-eight eggs, most of which were successfully hatched. Dump nests always contain traces of down and essentially clean eggs in a tidy nest.

Many dump nests are incubated when a dominant hen takes charge by driving the other hen or hens away. If a dump nest has twenty or more eggs, and the eggs are incubated, fewer than fifty percent of the eggs in the clutch typically hatch, compared with well over ninety percent that hatch in a normal clutch. Twenty or more eggs are usually too many for one small hen to keep uniformly moist, warm, and properly rotated. However, an occasional dump nest will produce twenty to thirty ducklings, and most studies of nestboxes confirm that dump nests produce a greater number of live ducklings than normal nests. A "drop nest" is similar to a dump nest except that the eggs are thought to be laid by yearling hens, abandoned, often quite soiled and discolored, and are never covered with either litter or down.

An essential factor for generating the energy required for such a high rate of egg production—the highest of any North American duck—is the food supply. During the winter prior to nesting, a female Wood Duck's diet consists primarily of various types of seeds, including acorns, beechnuts, and other forest mast as well as floating plants in areas where they are available, such as duckweed and water-fern, which she consumes to fortify herself for the rigors of winter. Fat thus produced and stored helps sustain her during the spring migration and contributes to the egg production soon to follow. In the days immediately preceding the laying of her first egg, the hen radically changes her diet to one dominated by invertebrates, insects, and occasionally frogs or

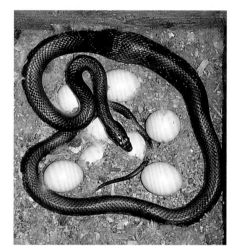

TOP The Wood Duck's diet, more varied than that of most other ducks, includes aquatic plants, seeds, and insects.

ABOVE, FROM LEFT Among the Wood Duck's many predators are Opossums, egg-eating snakes, and Great Horned Owls.

LEFT Wood Ducks occasionally feast on frogs, salamanders, and small fish.

BELOW Ducklings use a temporary "tooth," located on the tip of the bill, to cut their way out of the egg. This egg tooth is shed a day or two after the ducklings hatch.

BOTTOM Newly hatched ducklings are wet and exhausted, but within twenty-four hours they will be in the water chasing insects.

small fish, which provide the protein essential for egg production. Since protein cannot be stored in the body, it has been calculated that a hen must consume over five thousand insects, or the equivalent of one insect every five and one-half seconds for eight hours each day, in order to produce a single egg. During the laying period, she consumes a quantity of food roughly equivalent to her own body weight—within two weeks.

As a pair of Wood Ducks feed together during the laying period, the hen is constantly in motion, searching for insects, picking at every leaf and stem of grass, or probing in the mud along the edge of a pond or waterway. Meanwhile the well-fed drake swims leisurely by her side—disarmingly alert and on guard.

When the last egg is laid, the hen begins her solitary vigil deep within her dark nest. For approximately thirty days and thirty nights, without any direct assistance from the drake, she incubates her eggs. In natural nest cavities, which often have large entry holes, this is a particularly precarious time since many predators are able to gain easy access to the hen, virtually trapped in her own nest. Throughout the contiguous United States, the Raccoon is unquestionably the most formidable four-footed predator of Wood Ducks. Bobcats and Mink, agile swimmers and tree climbers, are able to invade many natural cavities or grab a hen as she is flushed from her tree hole. Rat snakes, both Gray and Black, take a terrible toll of eggs in the southern states. Natural cavities offer little protection against Sharp-shinned and Cooper's Hawks. Predation of natural nests is much more common than of nestboxes specifically designed to keep out most invaders.

During the twenty-eight to thirty-three days of incubation, after covering the eggs with her down, the hen usually leaves the nest early every morning and late each afternoon, occasionally even at night, to join her mate to feed and loaf. She feeds far more actively than the drake, since these short periods off the nest are her only opportunities to obtain nourishment. She may be away from the nest only an hour or so in the early morning and again in the evening, although on days when she has been disturbed, she may remain away for three to four hours. On one occasion, a hen we were photographing in California left her nest at approximately half past four in the afternoon and, apparently alarmed by our activities, had not yet returned by dark. After almost five hours, we were convinced she had abandoned the nest—or, if she did return, the eggs would have chilled and would not hatch—but even after such a prolonged absence, the eggs did hatch on schedule. In the early process of her incubation, a hen flushes more easily from the nest. As the time for hatching approaches, however, she generally remains protectively on her eggs, and may even peck or strike with her wing anyone attempting to lift her off the nest. The response of the nesting hens to human presence clearly demonstrates the dramatic differences in temperament among the birds. Some hens accept human intrusion from the first encounter. Other hens, at the first sign of danger, quietly leave the nest after carefully covering the eggs with their down to keep them warm during their absence. Still others bolt the nest, leaving the eggs uncovered and often soiled.

During the fourth week, usually two or three days before the chicks are hatched, there is noticeable activity within the shells. The first visible sign of life is the pipping or cracking of the eggshell in a circle at the large end of the egg. Holding the egg to an ear, one can hear the chipping of the sharp temporary egg tooth located atop the encased duckling's still soft and pliable beak. This calcareous tooth enables the tiny energetic duckling to emerge at last from its shell, having cut through the egg's tough membrane and, finally, the shell itself, a process taking from twenty-four to thirty-six hours. A day or two after the duckling hatches, the egg tooth, lacking any other function, is shed. Even before the chipping is audible, one can hear the soft peeping of the industrious duckling inside the egg. When undisturbed, the mother duck answers with a gently repeated call, often described as *kuk-kuk-kuk*. Through this process, known as imprinting, the hen establishes an enduring attachment to her young ducklings. The mother duck will later use this call to summon the chicks from the nest often high in a tree, to join her on the water below, and then to teach them to feed or to hide. The phenomenon of a hen's imprint on her ducklings is critical to their survival during their first few hours and days of life.

Years ago at our Indian Meadow Ranch in northern California, we had set up to photograph an incubating hen returning to her nest from her afternoon feeding, her eggs already pipping and ready to hatch. As she paused in flight before the nesthole, she was snatched in midair by a Cooper's Hawk, which forced her to the base of the tree, where she was left mortally wounded. Hoping to save the already pipping eggs, we took them several miles away and set all fourteen of them under another hen that had been incubating her own eggs for less than one week. We then removed this latter duck's eggs and added them to the nest of still another

TOP On the day that the ducklings are ready to leave the nest, the hen takes a last flight to survey the area and then returns to warm them.

ABOVE The hen briefly broods the ducklings to dry them and give them strength before they begin their exodus.

hen that had eggs of exactly the same age. We wondered if the first foster mother would care for ducklings that would hatch three weeks prematurely, and if she attempted to call them from the nest, whether they would recognize and respond to their foster mother's voice since they had certainly been firmly imprinted earlier by the natural mother they had lost to the hawk. Fortunately, everything went as hoped. Two days later, the foster hen successfully hatched and called all fourteen of her adopted ducklings down from the nest before leading them into the cattails. Three weeks later the second hen hatched her own eggs in addition to those we had given her. This man-made dump nest produced a total of twenty-four ducklings.

Nothing in the literature on imprinting explained exactly what took place in the first instance. However, it would seem that a completely new imprinting by the adopting hen had occurred in the last forty-eight hours of the ducklings' pipping, hatching, and departure from the nest. In spite of this successful imprinting, there is no assurance that all hens will accept such radically premature ducklings.

After pipping begins, the ducklings may take twenty-four to forty-eight hours to free themselves from their shells. Newly hatched ducklings are wet and exhausted and for the first few hours appear nearly lifeless, but dry quickly and rapidly gain strength. Wood Ducks are precocial, meaning that within a few hours after hatching, still covered only with down, they are capable of leaving the nest and surviving, with the assistance of the hen to brood them. However, unless the hen is disturbed, she usually will keep the ducklings on the nest for at least another twenty-four hours while they dry and grow stronger. On the day the ducklings are ready to leave the nest, the hen often takes a last morning flight, occasionally to feed and to scout the area, before returning to the nest. Newly hatched ducklings remain in the nest deep in the down and litter until, at last, when they are strong and restless, the hen calls them to join her on the water or ground below.

For a one-day-old duckling to join its softly calling mother, often twenty to fifty feet below the nest, requires the agility of an Olympic gymnast. To climb out of a nest in a natural cavity, a single chick may need to ascend vertically to the nesthole entrance, often a distance of forty-eight inches or more. To make the departure even more hectic, ten to twenty ducklings may be struggling to leave the nest at the same time. Their curved claws, sharper than those of any ground-nesting ducklings, are able to grip the wood much as a lineman wearing sharp spikes climbs a utility pole. The ducklings' climb from inside man-made nestboxes is made easier if the wood below the hole is roughed or scarified or fitted with material such as a strip of metal hardware cloth to act as a ladder. The ducklings accomplish this ascent by making a series of short upward jumps or hitches rather than actually climbing to reach what was the mother's entry hole to the nest and has now become the ducklings' escape hatch. No matter what the surface or height, the ascent and subsequent descent are exceptional feats.

Old stories tell of the mother duck carrying ducklings from the nest in her beak, under her wing, or on her back to a soft landing below the nest. Even such an authority as John James Audubon claimed to have seen the same sight. To this day, however, hundreds, perhaps thousands, of observers using binoculars, telescopes, and cameras have watched Wood Duck chicks leave the nest but none have seen the hen play any part other than to call them out to join her. This is a solo performance, and in only a minute or two, every able duckling, each resembling a small tuft of down, leaps from the nesthole, tiny wings spread to slow the fall, feet extended as if in expectation of a perfect landing. Few ducklings are ever injured by the fall no matter how

Just before summoning the ducklings from the nest, the hen scouts the surroundings, then flies to the water below. Hearing her soft call that all is safe, the ducklings scramble to the nesthole and, one by one, jump down to join her.

The hen leads her young ducklings to the protection of the covered shoreline. Of the ten ducklings that on average leave the nest, half are lost in the first two weeks of life. Only two or three ultimately survive—enough to support a thriving population.

great the distance. More ducklings are lost inside the nest, being too weak to climb to the exit hole, and are abandoned by the hen as she senses instinctively that the last able duckling has joined her in her urgent search for food and shelter. Most hens remain with the ducklings for the next eight to ten weeks until, as fledglings, they are ready to fly.

During the first few hours of a duckling's life, its downy covering lacks the ability to shed the often chilling waters into which the hen has led her brood. At Indian Meadow Ranch in California, we had to learn this the hard way. Mallards nesting on the ground on the islands in the ranch ponds seemed ideal foster parents to incubate the eggs we had taken from nesting Wood Ducks, in the hope of encouraging them to start a second nest. Several Mallard hens successfully hatched such adopted chicks, which could be observed following their foster mothers around the lakes. We soon realized, however, that after the first or second day, we never saw them together again.

This caused us to focus more closely on the next such brood to hatch. We watched the Mallard hen swim around the lake and finally climb out on the shore. The day-old brood of Wood Ducks scampered out to join her. The ducklings were obviously very wet and cold, and in need of warmth and shelter. As they approached the hen, she moved away a few steps—never leaving them, but never offering to brood them, as we had often seen Wood Duck hens cover their ducklings under their wings and body feathers. After several minutes, the Mallard hen disappeared from sight into the berry bush with the rejected chicks following after her. Two days later, this easily identifiable Mallard hen was on the island—but without the ducklings. We concluded that, for reasons we did not understand at the time, Mallard hens on the ranch ponds could not successfully raise young Wood Ducks.

What we now know is that Wood Ducks prepare their small ducklings for the long leap from the nest to the land or water below by brooding them, letting them dry and gain strength, for at least twenty-four hours, and occasionally for as long as thirty-six hours, before calling them out of the nest. By contrast, Mallards at the ranch take their ducklings into the water within less than twelve hours after hatching, and the hen did the same with the adopted Wood Ducks. Thus, quite unknowingly, the Mallard exposed the foster ducklings to the cold water of early spring some twelve or more hours sooner than nature had prepared them for it, yet she would not accept the prematurely chilled ducklings under the protective warmth of her wings. We can only assume that perhaps voice imprinting between the Mallard and the young Wood Ducks was inadequate.

The apparent failure of imprinting between species with entirely different habits is addressed by Ray Cunningham of Minnesota, who wrote in the *Wood Duck Newsgram* of August 1991: "Regarding wood ducks and hooded mergansers, my experience indicates that wood duck ducklings do not respond to calls of a hooded merganser hen, even though they may have been hatched by her, nor do hooded ducklings respond to wood duck hens in a similar situation." In any event, hatching ducklings under other species of ducks seemed too risky to try again.

The best nesting sites for Wood Ducks are often in densely wooded swamplands or riverbank areas. Yet the swamplands frequently lack the sunlight, and rivers the shallow, slow-moving water, to produce an adequate food base. Thus the area where a Wood Duck nests and where her ducklings hatch is frequently not the best habitat for raising the ducklings. In this early stage of their lives, ducklings require a diet with maximum quantities of protein, primarily in the form of insects and other invertebrates living in floating vegetation, the surrounding grasses, and other plants. There are many tales of hens leading their broods more than a mile from the nest in search of food. Dr. F. Eugene Hester and Jack Dermid, writing of brood

Day-old ducklings are precocial—able to leave the nest and fend for themselves. Songbirds, which are altricial, require weeks of care before they can leave the nest.

When this molting drake loses its flight feathers, it will need to seek protection from ever-present predators. The severe molt cannot obscure the clean sculpting of the Wood Duck's bill to its head—the broad notch extending upward almost even with the eye.

behavior in South Carolina, tell of five separate broods that were tracked from their nesting area to a millpond a mile and a half away. All five broods reached the millpond within eight to forty-seven hours. Nests are rarely located more than a mile or so from water, for ducklings must find their first food within forty-eight hours of hatching—and a mile can be a long way for day-old ducklings to travel.

In northern California, we have often watched young Wood Duck ducklings follow their hen down the falls below the main lake and into Barnes Creek. By nightfall, we lose sight of them, but by the next day, we have spotted them almost two miles away, still traveling among the boulders in the creek, nibbling everything edible along the route. By the third day, we are never able to locate them and hope that they have reached the shelter of the larger Mayacama Creek and, finally, the Russian River almost three miles away. The farther a brood has to follow the hen from her nest to proper feeding grounds, the greater the loss of life to predators.

For thirty-five years, Frederic Leopold, the youngest brother of Aldo Leopold, studied and wrote about his nesthouses for Wood Ducks on a bluff in Burlington, Iowa, high above the Mississippi River, yet little more than a stone's throw across the railroad tracks and to the river itself. He reported that in one year, of the one hundred eighty-nine ducklings that attempted to follow their hens on the short but precipitous hike from the bluff to the river below, sixty-one became lost or entangled in the dense brush and brambles along the way and, for that reason alone, never survived the journey. Unfortunately, nearly half of all Wood Duck ducklings hatched throughout the country are lost during the first two to three weeks of life.

SUMMER Of all the seasons, summer is, in some ways, the most hazardous and difficult for the Wood Duck hen. The hen, already deserted by the drake for several weeks, alone tends her newly hatched and vulnerable brood against the elements and guards them from the devastating attacks of predators for most of the next six to eight weeks until the ducklings are able to care for themselves. The drake is already in his early summer, or eclipse, molt. Even while incubating her eggs, the hen, too, will slowly start a molt that will continue until her ducklings are ready to fly.

Arthur Cleveland Bent's colorful description of

BELOW An immature duck, more than three months old and already strong of wing, shows the characteristic white fingermarks of a drake on the side of the head.
BOTTOM Rapidly maturing juveniles wend their way through water lilies.

Molting drakes often congregate far from their nesting sites as they shed their nuptial plumage.

the Wood Duck drake as a "Beau Brummel among birds" is appropriate, not only for the drake's elegant appearance, but for his itinerant and bachelorlike propensities as well. Except for a brief courtship, primarily in winter and early spring, the drake shows only sporadic interest in his hen and none at all in their offspring. He plays little part in the choice of a nest site, provides no assistance, as many other birds do, in incubating the eggs, and deserts his mate a week or more before the eggs are pipping and the ducklings are hatched. Many drakes congregate together in densely covered wetlands often far from the nesting area to begin their summer molt, during which they lose their brilliant nuptial plumage and assume the eclipse plumage that is worn usually from early May through August. The drakes now look more like hens in their normal plumage or even like the inconspicuous, but rapidly maturing, juveniles.

Early in their summer molt, the drakes, in a manner common to many species of waterfowl, enter the so-called remigial molt, the loss of all flight feathers, which leaves the drakes flightless, normally for twenty-two to twenty-four days. During this time of extreme vulnerability to predators, the drakes are so secretive that few can be observed even when an intensive search is made. Many flightless birds seek sanctuary in areas inaccessible to humans. It is crucial to the survival of the species that the drake, having molted his splendid nuptial plumage, is drab and colorless and, therefore, better able to avoid detection by predators of all kinds.

Summer is an equally perilous time for those hens and their ducklings that have survived the sometimes long overland trek to find a sheltered brooding area on rivers and ponds filled with emergent pondweed containing abundant sources of protein. In addition to Raccoons, Bobcats, foxes, and other four-footed predators, hawks and owls pursue the brood. Once the newborn ducklings are on the water, the most dangerous predators are Largemouth Bass, pike, and pickerel, and the equally voracious Bullfrogs and snapping turtles lurking in the dark water below. There is simply no protection against their silent attack. It is during the first weeks of the ducklings' life that predators of all kinds find them an easy target. Of the fifty percent of all ducklings that perish before they can fly, seventy-five percent of those are lost during the first week after leaving the nest.

BELOW In the fall, Woodies normally spend the night roosting on the water in habitat that provides dense and sheltering cover.
BOTTOM At eight to ten weeks of age, the ducks will be able to fly.

The nature of brooding habitat varies tremendously from area to area. The most exposed habitats are the most susceptible to predation, whereas ponds more heavily interlaced with fallen logs, patches of cattail, and islands of willows, Buttonbush, or other growth offer better protection and are able to sustain a greater number of broods. A large open pond or swamp may support only a single brood, but a very small, heavily overgrown pond may contain several broods. Dr. F. Eugene Hester and Jack Dermid cite their own experience in North Carolina on a pond with heavy growth, where nineteen broods were reared on under four acres. Frank Bellrose observed thirty-four broods on fifteen acres on Nauvoo Slough in Illinois. The protective nature of the habitat seems more important than its size, always assuming an adequate food supply is available. A single acre of brooding area under optimum conditions is thought to be capable of supporting ten to fifteen broods, totaling perhaps a hundred ducklings, without exhausting the proteins and carbohydrates they require. For so many ducklings to survive, that acre of water also should have a heavily wooded shoreline and should include numerous areas of shrubs and brush offering refuge from predators and places for sunning and preening.

Across the North American continent,

the seven to eight million Wood Ducks that survive until fall migration are reduced, primarily by natural causes and hunting, to approximately three million ducks by early spring. Less than half this number are nesting females. More than half of these fail, for many reasons including predation and storms, to hatch even one duckling. However, many hens that survive the loss of their first nest usually will nest a second time. The net result is that during spring and early summer an astounding six to eight million ducklings tumble from their nests full of energy and expectation. By the time the average brood of ten ducklings reaches flight stage, fewer than five survive to fly.

Except for the Asian Mandarin, no other ducks are able to raise so many ducklings to flight stage. This high survival rate is possible for several reasons: The Wood Duck lays larger clutches of eggs than most other ducks, and the percentage of fertile eggs, ninety to ninety-eight percent, is unusually high for any species of duck. Also contributing to the high recruitment rate is the rare ability of Wood Duck hens to raise two successful broods in a single season. This is most common in the Wood Duck's southern ranges, where the breeding season is longer and the food supply is greater. Since most drakes in these areas have entered their eclipse molt earlier in the season, and are generally considered infertile during their molt, it is unclear how a second nesting hen finds an effective mate. In spite of this, the melancholy call of a female seeking a second mate can be heard traveling eerily across the waters and through the trees, and quite evidently succeeds in attracting a virile mate—perhaps an unpaired yearling or adult male not yet into his molt.

One hen, for instance, observed over a period of several years by Stephen Simmons in central California near Merced, each year started laying her second clutch of eggs while the chicks in her first brood were only twenty-seven days old, less than halfway to flight stage. Since it is highly unlikely that she lost her entire first brood each of those years, perhaps she had simply left them to fend for themselves. At that age young Wood Ducks, more than other ducks, can survive on their own. Ducklings abandoned while too young might even be adopted and reared by a second hen. This unusual characteristic of Wood Duck hens to adopt deserted ducklings is still another factor in the successful recruitment rate of the species. In our experiences with second nestings in northern California, a second clutch generally has fewer eggs than the first, and the rate of fertility is lower. By contrast, Stephen Simmons, working less than two hundred miles south of us and on a much larger scale, found that the second clutch contains an equal number of eggs with the same degree of fertility as the first clutch. This is obviously the result of a high energy level in the hen

By midautumn, the Wood Ducks have regained their colorful plumage and are ready for fall migration.

and possibly in the drake, which means that the lush food supply in Simmons's area sustains both hen and drake far better than the more restricted supply farther north.

With the passing of summer and prior to fall migration, Wood Ducks, in an unusual activity, gather together in flocks at night. Such nocturnal roosting, unlike roosts of songbirds on perches or in trees, takes place on the waters of lakes and ponds, beginning shortly before sunset and continuing until dark. Hundreds of Woodies congregate in a single roost, and as the first ponds and swamps begin to freeze, the concentration of ducks on open water continues to increase. Autumn flocks of Wood Ducks in nocturnal roosts have attracted great attention since, at this time of the year, small groups of ducks are rarely seen. The sight and sound of thousands of Wood Ducks flying to their nocturnal retreats is an awe-inspiring experience.

By early fall, four to five million young Wood Ducks will have survived to join almost an equal number of adult birds. Together they face the relative serenity of early autumn and the harsh vicissitudes of migration and the long winter that will reduce their numbers by half—as once again the season turns full circle.　　　　　　　　　　　　　　　　　　　　　　　　[L.L.S.]

CHAPTER FOUR

THE WORLD OF THE MANDARIN

W ITH GREAT NOSTALGIA I READ THESE LINES, which I wrote over forty years ago in the preface to my book *The Mandarin Duck*. My introduction to the Mandarin took place in England, where they, while not native to this land, have been free-flying for nearly sixty years. Watching the Mandarin on Virginia Water, a beautiful lake in southern England, I came to know and admire them as the most beautiful of all ducks, some say the most beautiful of all birds.

For the past ten years, however, it has been my good fortune to observe the Mandarin in an important part of their natural habitat in my temporary home in Japan, which is in the southeastern portion of their flyway in East Asia. The entire flyway extends southward from the Mandarin's spring nesting grounds in Ussuriland in the extreme southeastern part of Siberia to what appear to be ever-dwindling nesting and wintering grounds in the Koreas and eastern, central, and even southern China, and inland to Myanmar (Burma) and northeast India. Small numbers of birds have occasionally been sighted in Vietnam, Thailand, and Nepal. Reaching east across the Sea of Japan, the flyway encompasses some of the Mandarin's most important wintering and nesting locations in the islands of Japan, including Hokkaido in the north, and extends as far south as Taiwan.

The Mandarin is most commonly said to be a resident of the Far East. This has been the perspective of most writers residing in Europe or the United States, who see Asia as located far to the east, on the other side of the world.

It often happens that we find the most beautiful things in life when they are least expected. So it was with me when first I saw the Mandarin ducks at Virginia Water. I had come upon them unaware in a secluded creek, guarded by banks of rhododendrons. They were away on the wing in a flash, but their fantastic form and color had captured my fancy.
—CHRISTOPHER SAVAGE

OPPOSITE The Mandarin, the "pearl of Manchuria," is known as the Mandarinka in Russia, the Yuen Yang in China, and the Oshidori in Japan.

115

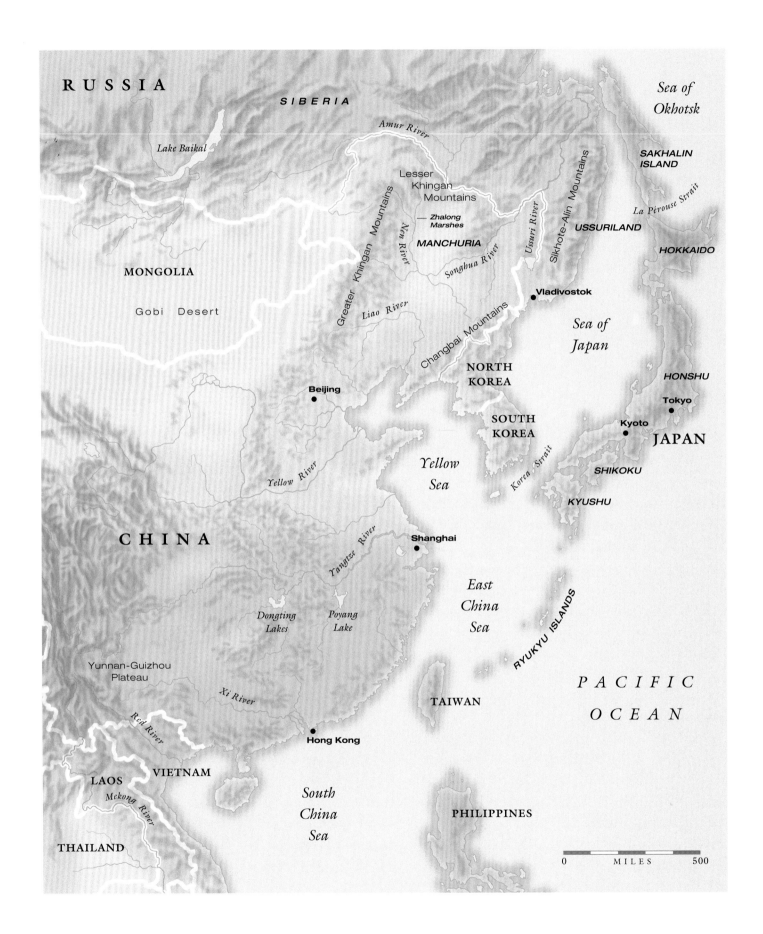

RUSSIA

SIBERIA

Sea of Okhotsk

Lake Baikal

Amur River

Lesser Khingan Mountains

SAKHALIN ISLAND

— *Zhalong Marshes*

La Pérouse Strait

Ussuri River

Sikhote-Alin Mountains

USSURILAND

HOKKAIDO

Nen River

MANCHURIA

Greater Khingan Mountains

MONGOLIA

Gobi   Desert

Songhua River

Liao River

Changbai Mountains

Vladivostok

Sea of Japan

NORTH KOREA

HONSHU

Tokyo

Kyoto

Beijing

SOUTH KOREA

JAPAN

Korea Strait

SHIKOKU

C H I N A

Yellow River

Yellow Sea

KYUSHU

Yangtze River

Shanghai

Dongting Lakes

Poyang Lake

East China Sea

RYUKYU ISLANDS

Yunnan-Guizhou Plateau

Xi River

PACIFIC OCEAN

Red River

Hong Kong

TAIWAN

LAOS

VIETNAM

Mekong River

South China Sea

PHILIPPINES

THAILAND

0   M I L E S   500

From my vantage point in Japan, and because of my interest in the Wood Duck of North America as well as the Mandarin, I was fascinated by a review by Dr. Miklos D. F. Udvardy of a book published in 1990 in Germany, titled *Die Vogelwelt Ussuriens (The Birds of Ussuriland)*, by Algirdas Knystautas and Dr. Yuri Shibnev, describing wildlife in Ussuriland, a narrow strip of land in the southeasternmost portion of Russia on the Sea of Japan.

Dr. Udvardy is a well-known ornithologist, biogeographer, and author. One of his most popular publications is *The Audubon Society Field Guide to North American Birds (Western Region)*. His biogeographical background gave his review a unique perspective. Dr. Udvardy notes that the harbor city of Vladivostok in Ussuriland is located at precisely the same latitude as Eugene, Oregon, eastward across the Pacific Ocean, in the United States. He compares the two landscapes and the creatures that live on opposite sides of that great sea. Each coast is flanked by an island—North America by Vancouver Island and East Asia by Sakhalin Island. Each consists of great mountain ranges punctuated by river valleys—the Columbia in Oregon and the Ussuri and Amur in Ussuriland.

Without intending to do so, Dr. Udvardy was describing precisely the northern nesting grounds of the Mandarin in East Asia and those of the Wood Duck on the Pacific Flyway of North America, not from a European or North American perspective, but from a point centered in the Pacific Ocean (see map, pages 18–19). In concluding his comparison of the landscapes of Asia and North America, Udvardy tells us that the Siberian taiga, or coniferous forests, could be the "mother ecosystem" for the northern mixed and coniferous forests of the North American continent, and that the Mandarin and the Wood Duck are examples of the very few related species inhabiting both regions. This suggests the romantic possibility that just as the earliest humans are thought to have crossed the Bering "ice" bridge from Asia to North America along with, or in pursuit of, ancient mammals, a common ancestor of the Mandarin and the Wood Duck might have journeyed across the same bridge, thus explaining the existence of these two similar species on opposite shores of the world's largest ocean.

Whereas the total landmass of Asia is many times larger than that of North America, the peculiarities of climate and habitat limit the Mandarin to an extremely narrow eastern coastal area and the islands adjoining it between approximately twenty-five and fifty-five degrees north latitude—similar to the Wood Duck's Pacific Flyway, which extends from approximately thirty to fifty degrees north latitude. In addition to its presence on the Pacific Flyway, the Wood Duck is found throughout much of the North American continent, where its total population in autumn is estimated at seven to eight million

OPPOSITE The Mandarin's flyway in Asia, on the western shore of the Pacific Ocean, covers nearly one million square miles—only a fraction of the entire continent. Stretching over two thousand miles north to south, it ranges from southeasternmost Siberia in the north through the coastal region of China and the Koreas, east to the Japanese archipelago, and south almost to Hong Kong.

birds. Thus, despite the great size of the two continents, a distinction must be made between the Mandarin's limited population potential in its restricted Asian habitat and the large population of Wood Ducks occupying the great center of the North American continent. A more appropriate comparison would be between the population potential for the Mandarin in its flyway and the population in the fall of approximately one hundred thirty thousand Wood Ducks on the Pacific Flyway. The population of Mandarin in East Asia today is far less than that and is thought to be decreasing.

Since the time of Confucius, the Mandarin Duck has been celebrated in Asian art and literature as a symbol of fidelity and devotion. Paradoxically, serious writing on its natural history originated in the West—in England, Germany, and the United States. As a result, relatively little is known about the isolated world that the Mandarin inhabits or its movements within that world. This lack of precise documentation is understandable. The Mandarin's domain crosses the political boundaries of countries whose people for generations have scarcely communicated with one another and, except in times of war, have had little access to each other's territories.

The Mandarin, like the Wood Duck, is believed to be semimigratory and semicolonial. Almost no ringing, or banding, of Mandarin has ever been attempted, and hunting reports are virtually nonexistent since the hunting of Mandarin is illegal throughout its territory. Nevertheless, aided by scarce bits and pieces of evidence, it is possible to speculate on the Mandarin's migratory routes, its nesting and wintering grounds, the areas where it may be resident year-round, and the status of its habitat, seemingly so severely diminished by agriculture, deforestation, and pollution. To discover more about the Mandarin, I have relied not only on my own experience but also on the latest information available from the countries the bird inhabits, mainly Russia, China, and Japan. For a species so widely admired throughout the western world—and so venerated in the religion and art of much of Asia—too little is known of the Mandarin's current status.

OPPOSITE Ussuriland—mountain-ous, rugged, and isolated—is the most northerly nesting area of the Mandarin.

RUSSIAN USSURILAND

The remote wilderness known to Russians as Ussuriland—the northernmost range of the Mandarin in the Far East—is a land both fascinating and forbidding. Within this relatively narrow portion of southeastern Siberia—extending approximately six hundred miles north to south and two hundred miles east to west—wildlife can be found in abundance, amid dramatic contrasts of terrain. The main feature that defines the rugged character of Ussuriland is the four- to six-thousand-foot-high Sikhote-Alin mountain range running its full length. The eastern side drops off precipitously toward the Sea of Japan. The gentler western slope eventually meets the massive Amur and Ussuri Rivers, which separate Ussuriland from Siberia to the north and Manchuria and North Korea to the south.

This northern land has always intrigued explorers and naturalists who were prepared to endure its formidable climate and terrain. The esteemed Russian ornithologist Vladimir E. Flint has traveled on foot through the forests of Ussuriland, which he describes as so dense that packhorses could barely pass through them. When the heat is not stifling—and the air not filled with clouds

Range of the Mandarin in Ussuriland.

of mosquitoes and gnats—there are furious storms that leave fallen trees to create almost impenetrable entanglements. To study wildlife here, Flint writes that one needs not only great enthusiasm but also an iron constitution, perseverance, and courage.

The wildness and isolation of Ussuriland are reflected in the lives of its people. Dr. Barry Hughes of the Wildfowl and Wetlands Trust in England and Dr. David Bell traveled in the early 1990s to the Bikin River basin and adjacent areas on the west side of the Sikhote-Alin Mountains to study the rare Chinese, or Scaly-sided, Merganser and to assess the damage to its nesting areas by agriculture, mining, and the heavy harvesting of timber. They describe hunting, trapping, and fishing as the major subsistence activities. In the harsh winters of the hunting and trapping season, temperatures reach minus-thirty degrees Fahrenheit. Lakes and streams are frozen solid and snow covers the landscape. The deer that the hunters kill and cannot consume are preserved by smoking. Sable and badger pelts are sold to provide cash for other supplies. During the winter, the fur-clad hunters live for months at a time isolated in hamlets consisting of a few wooden huts with thatched roofs and small glass windows that keep out the cold but let little light into the dark interiors. Their families live many miles away in small villages near the Ussuri River, with electricity and schooling for their children. As the winter eases, the men begin to fish in the Bikin and other major rivers or their tributaries. It is a primitive and lonely life.

TOP During the Siberian winters, every pond and stream freezes, driving Mandarin far south to await the spring.

ABOVE Native trappers and fishermen describe their lives and the Ussuri wildlife to researchers from England, China, and Russia.

Those outsiders adventurous enough to explore this challenging land are rewarded with an experience of remote mountains, majestic woods, and clear-running streams inhabited by exotic wildlife. Of nearly eight hundred avian species known to live throughout the former U.S.S.R., four hundred are found in Ussuriland, among such extraordinary mammals as the Siberian Tiger, Asian Black Bear, rare Amur Leopard, and Sika Deer. One of these avian species is the Mandarin Duck—known in Russia as the Mandarinka—described by Vladimir Flint as the "pearl of Manchuria."

To this wilderness so inviting to wildlife but so hostile to humans, the Mandarin arrive in early spring. What brings these birds so far north each year? What is special about this narrow stretch of land, out of the eight and a half million square miles of the former U.S.S.R. stretching westward some seven thousand miles and spanning eleven time zones? Why do they choose this harsh part of the world for spring nesting and raising their broods? Why do they spend half their lives in this forbidding land?

In all of Russia, only Ussuriland provides the habitat the Mandarin needs to survive. And of all the areas on the Mandarin's Asian flyway, Ussuriland offers less competition for the species' most essential necessities: the special tree cavities without which they are unable to reproduce, and the availability of fresh water and acorns and other food without which they are unable to survive and raise their young. No reliable estimates have been made of the percentage of Mandarin from Japan, China, the Koreas, and other southerly areas in the flyway that in spring migrate north to Ussuriland. But it may be relevant that approximately one-third of all North American Wood Ducks are believed to remain near their wintering grounds, while the other two-thirds fly north to breed and nest. It is entirely possible, therefore, that most East Asian Mandarin will nest in the woodlands and wetlands of Ussuriland and northern China.

The Ussuri wilderness, where Mandarin nest in spring, is a welcome contrast to their wintering grounds near the areas of China with the most dense human population in the world.

In much of China and on the islands of Japan, the large and ancient trees that provide natural cavities and adequate food supplies are becoming ever more difficult for the Mandarin to find, not because the species is growing in numbers, but because its habitat continues to shrink with the increasing encroachment of civilization. Ussuriland, in its relatively primitive state, may be a nesting area of last resort. The highest elevations of the Sikhote-Alin Mountains are covered almost exclusively with a dense growth of conifers, often referred to in the subarctic regions as *taiga*. Conifers are mostly anathema to Mandarin since they rarely provide suitably large nesting hollows, nor do they supply the seeds or fruit Mandarin need to survive. Mandarin thrive in the mixed and broad-leaved deciduous forests that abound in the valleys just below the taiga. On the inland side of the Sikhote-Alin, the valleys descend and widen into the vast Amur and Ussuri river basins, where the streams are placid. On the Pacific side of the range, where the mountains are much closer to the sea, the valleys are shorter and the rivers more turbulent.

The slopes of these valleys are dominated by giant Mongolian Oaks, the Mandarin's tree of choice, offering many of the best nesting cavities and the acorns that constitute a major part of its diet. The forest also contains an abundance of willow, poplar, elm, apple, and birch trees, and Amur Lilac. In most cases, the ducks prefer the shallow waters of the inland streams with their many sandbars and quiet backwaters. However, Mandarin, unlike the North American Wood Duck, seem equally at home in slightly faster running streams, and can often be found on the rocks in streams that tumble through narrow channels. The rivers and lakes of Ussuriland also provide the invertebrates, insects, and amphibians that play an important role in the Mandarin's diet. These are the qualities of life the Mandarin needs to survive, and these are what Ussuriland, for all its other more rugged and forbidding characteristics, has to offer during the six to eight months of the birds' nesting and brood-rearing season.

Early adventurers and explorers courageous and determined enough to reach the most remote areas of Ussuriland encountered Mandarin —the "pearl of Manchuria" — in the pools of the tumbling mountain streams.

SPRING MIGRATION Much of what we know of Mandarin in Ussuriland has been reported by a pioneering group of fewer than a dozen Russian scientists from their travels through the area. The observations of many of them were recounted at a symposium in Vladivostok in 1985 in a collection of scientific papers entitled *Rare and Endangered Birds of the Far East*. They describe Mandarin arriving from the southern parts of the flyway as early as the first week in March, near the southern seashore, where the Pacific currents warm the frozen land. Farther north, the Straits of Tatar between Sakhalin Island and the mainland remain frozen. In the spring, even more southerly lakes show only potholes of open water in the ice. Most Mandarin are reported to reach Ussuriland, even the southern parts of Sakhalin Island, in the first two weeks of April. Early in migration, birds often arrive in groups of four to six individuals, and at other times only solitary pairs fly in. Later in spring the birds move farther north past Khabarovsk, an isolated industrial city on the Trans-Siberian Railroad, near the juncture of the Ussuri and Amur Rivers. Still farther north, toward the mouth of the Amur, groups of up to a dozen arrive, and sometimes as many as forty drop to the lower, warmer reaches of the rivers and their tributaries until the end of April, when larger numbers pour in to lakes and rivers the full length of Ussuriland and even to its seacoast.

Mandarin, like North American Wood Ducks, court and mate in the water.

Most Mandarin fly in already paired for spring nesting. With males consistently outnumbering females, however, a familiar sight is of several males pursuing a single female. Such a trio of two males and one female fed for a long time on a sandbar near the bank of a river in the Bikin River basin. There was little activity until one male casually swam toward the lone female, whistling his faint call, and after a few displays of mutual courtship, a brief mating took

There may be more natural nest cavities in Ussuriland than in all the other regions on the Mandarin's two-thousand-mile-long flyway. The Mandarin's unusually large and sensitive eyes enable the female to locate even the most inconspicuous nest sites.

place—a rather unromantic courtship for birds famous for their deep devotion to their mates. Finally the three birds continued to feed and rest until it was too dark to observe them longer. By early May, all migrating birds have arrived, and nesting will soon begin.

NEST SEARCH Immediately following their arrival, the Mandarin start the search for suitable nest sites, which are large, deep cavities found primarily in old oaks, willows, or other broad-leaved deciduous trees. The hen usually leads the search, although at times the drake seems to be urging his mate toward a particular site. In Ussuriland, the females often select nests close to slow-running streams or in the slightly marshy areas at the mouths of the small tributaries of the larger rivers. Frequently Mandarin nest near primitive roadways or beside brooks surrounded by agricultural land. When suitable sites are scarce, Mandarin have been known, in rare instances, to nest in holes along a riverbank or under broken limbs of old trees.

On one of the tree-studded islands of the Bikin River, west of the Sikhote-Alin range, a pair was observed to settle in the branches of a huge elm, then to fly to a poplar, stretching their necks and peering about, after which the male led the way to another hollow. He fluttered about the tree as if inviting the female to inspect it. He repeated this routine until the female flew over and, clinging to the tree like a woodpecker, looked into the cavity before flying away. The drake continued to invite her toward the prospective nest site, and only after she had looked into it once again did they depart together, to feed in a nearby channel. They continued to remain in the vicinity, and two weeks later, the female was found setting on six eggs in that same hollow.

The choice of the nest site is thought to be influenced more by the suitability and comfort of the hollow itself than by the qualities of the surrounding habitat. Females frequently nest in the same tree year after year. In fact, most hens that have successfully hatched their ducklings the prior year seek the same nest again, and their female offspring often use nests nearby. A local farmer tells of one nest in an elm tree standing in the middle of a hay field that had been used for seven consecutive years. It was abandoned only after someone, or something, had destroyed the entry hole.

NESTING   For most birds, once the nest site has been selected, the next task, and for many an arduous and demanding one, is the construction of the nest. Not so for female Mandarin, who, like Wood Ducks, carry no nesting material to the cavity. Nests are normally chosen only if a cushion of leaves and wood chips is present with which to cover the first eggs.

Most Mandarin lay the first of their buff-colored eggs within a matter of days after the nest has been selected. Morning after morning, for six or seven days, the hen lays an additional egg, covering it and the others with whatever litter is already present in the hollow. On the sixth or seventh day, the female begins to shed or pull her own downy breast feathers to cover the eggs. This natural insulation allows the female to be away from the nest for an hour or more as she joins her mate for feeding and resting. Typically, the female does not begin incubation until the last egg is laid. The male plays no part in the incubation process, but stays near the nest and accompanies the female when she leaves the nest early in the morning or in the evening to feed in close-by channels and small pools. The female feeds usually in the early morning and late afternoon and less frequently after dark.

Mandarin drakes are among the few ducks who guard their hens and ducklings, often until the ducklings are able to fly. Despite such vigilance, half or more of the ducklings, out of an average brood of eight to ten, fail to survive the first two weeks of life.

The nests observed by the Russian scientists reporting from Ussuriland were in natural cavities, and most were from twenty to sixty feet above the ground. Eggs were arranged in one layer, and the larger the cavity, the greater number of eggs it seemed to contain, up to a maximum of fourteen. In the smaller nest cavities, the eggs numbered as few as seven. Despite the small sampling involved, this egg count is, for some reason, considerably less than for the free-flying Mandarin in England and the United States, where the average number of eggs per clutch is twelve to fourteen. Also, the number of ducklings in broods reported from Ussuriland is smaller than the number in broods in England and the United States, which tends to confirm the existence of smaller clutches of eggs.

On her precious clutch the female sits alone, night and day, for approximately thirty days, fearful of predators, hunching down with each crack of a twig or swish of wings in the night. At long last, the eggs hatch, and the chicks tumble to the ground and follow the hen to explore Ussuriland's many waterways, looking for food and for shelter from predators. Despite having found one nest with ducklings hatching, and another brood just emerged from its nest, no Russian scientist has ever reported seeing one of the most fascinating of all Mandarin activities—the ducklings leaving their nest, when, in almost the blink of an eye, tiny, golden balls of fluff will tumble from the tree hole to join the softly calling hen.

Ever alert to potential danger, the hen will quickly lead her ducklings deeper into the safety of the sheltering shoreline.

THE BROOD Newly hatched ducklings are extremely difficult to observe, for the nervously alert females keep them well hidden under the overhanging branches along the shore or in the streams. At the first sign or sound of danger, the young ducklings dive underwater until they reach the riverbank, where they conceal themselves in the long grass.

It is not unusual to find a hen and her ducklings more than a mile from where they are believed to have hatched. The hen's immediate concern is to find adequate food and shelter for her ducklings. During the first weeks of their lives, her active ducklings require the high protein of insects and invertebrates, and she will travel, as long as the ducklings are able, to find them.

The downy ducklings feed in puddles along the bank of a stream, pecking at insects in the weeds. The females often do not feed, but stand sentry only a short distance from the brood. When the female senses danger, she leads her brood to hide in the overgrown willows. By contrast, older ducklings and their mother flee into the water and, skimming along the surface, make their way downstream. Having traveled a safe distance, they may jump out on the bank as the female again remains watchful to be able to divert any intruder. Other observations indicate that this behavior is characteristic of the Mandarin duck: Downy ducklings climb onto a bank to hide beside their mother, while partly fledged ducklings dash away into the water, as the female, feigning injury, attempts to lure the predator away.

The majority of the young birds fortunate enough to survive the first precarious weeks of life are able to fly by August, in time for the autumn migration to more southerly wintering grounds. The female's often lonely job is done. Now should be her time for preening and resting, for enjoying the last weeks before the long and perilous journey south. But nature is not so generous.

Secure at last, she allows the ducklings to rest on her back—a rare sight among Mandarin hens and their broods.

MOLT In Ussuriland the female normally begins her molt just as her young are ready to take to the air, leaving her on her own for the first time since she laid her first egg nearly four months ago. Within a few weeks, however, her freedom is completely frustrated as, like other ducks, she loses all her flight feathers. Then, late in the summer, for some three to four weeks she is unable to fly and remains virtually helpless. Seeking safety from predators on land, from the air, and in the water, she finds shelter in the shadows of the forests.

If the female's molt were influenced only by the season, it would start, as does the drake's molt, in early summer and be completed in late September or early October. But for nesting hens, the obligations of motherhood override the dictates of the season by delaying her major molt until her young are able to fly. Single or nonnesting hens, however, molt earlier, as do the drakes, which at this time of year are often found in small groups, having moved into deeper shadows to complete their molt. By July in Ussuriland, drakes in brilliant plumage and bright sails are not seen at all. Instead, they are in their drab and ragged eclipse molt and are scarcely distinguishable from the hens.

During their summer molt, both sexes seek the protection of forested banks or areas where heavy branches overhang the water. It is fortunate that, while the drakes are flightless and more vulnerable to predators, their loss of color serves as camouflage. Then, by early fall, their molts complete, the drakes are in their sparkling nuptial plumage, their copper-colored sails shining, and the females reclaim their more subtle beauty.

PREDATORS The predators that threaten the adult Mandarin and their eggs or ducklings vary from one part of the Mandarin's range to another. Just north of Vladivostok, a Mandarin was observed attempting to defend itself against a Mink. In parts of the Far East a predator of the Mandarin is the Raccoon Dog, a member of the canine family resembling the North American Raccoon, a relative of the bear family. Although both species have the striped "mask" of a bandit, the Raccoon Dog has a short tail, and the Raccoon has a long tail ringed in brown. Both species swim well, climb trees, and are major predators of small animals and birds. Otters and polecats in Ussuriland also prey on Mandarin and their broods. The remains of four Mandarin were found near an Eagle Owl's nest. Grass snakes will enter Mandarin nests to consume their eggs.

The Mandarin's most persistent enemies in Ussuriland are loggers, hunters, and, in particular, poachers. Although shooting Mandarin is illegal, most hunters are unable to recognize them in flight and mistake them for other legally hunted species of ducks. However, the greatest human threat is not directly to the birds themselves but to their habitat. Sadly, the rich Ussuriland forests are being heavily harvested, resulting in the loss of nest sites and food sources. Clearing the forested channels for timber rafting has opened these wild areas to motorboats and poachers. The lower elevations of Ussuriland are being developed for farming, logging, and mining, which seriously threaten the otherwise pristine environment.

Nature reserves, or *zapovedniks,* offer the best safeguards for the Mandarin, as well as for other indigenous flora and fauna. One hundred fifty reserves throughout Russia, encompassing almost sixty thousand square miles, represent the best hope for preserving endangered species. In Ussuriland, a dozen such reserves provide a measure of protection for the Mandarin and the even more endangered, tree-nesting Chinese Merganser.

OPPOSITE A Mandarin drake, like other dabbling ducks, tips for acorns in shallow water.

LEFT Mandarin in the Far East consume more tadpoles (far left), frogs (center), and roe than Wood Ducks in North America, perhaps because acorns, like those from the Mongolian Oak (near left), their favorite food, are less plentiful.

FOOD The Mandarin's diet changes with the season. Dragonflies, dew worms, grasshoppers, small fish, frogs, mollusks, and even small snakes constitute the primary food supply in the summer, prior to early fall when acorns are available. During the nesting season in the Bikin River basin, Mandarin on pools within the forest dive to the bottom for small invertebrates and frogs. Between feeding periods, they retire to the trees to preen and polish their feathers, and to rest, finally tucking their heads halfway under their wings in wary relaxation. One Mandarin pair, feeding in a shallow lake for several hours, was observed catching fish and, most astonishingly, capturing and eating two huge frogs. On another occasion, two drakes were found entangled in a farmer's fishing line, where they had been chasing live fish, some of which were being used for bait. Unlike the North American Wood Ducks, which derive their protein primarily from insects both terrestrial and subaqueous, the Mandarin in Ussuriland, in their search for protein, seem to eat more amphibians and fish.

By September and October, on both sides of the Sikhote-Alin range, ripening acorns are the main staple in a Mandarin's diet. As soon as the fruit appears, the Mandarin drift into the forest to forage. Flocks dive for acorns under the oaks that overhang the edges of the stream. The muddy bottoms of the rivers become plowed up by diving or tipping birds in search of acorns, roots, and other foodstuff. Oftentimes the ducks fly to the trees and pluck the acorns off the branches.

Mandarin also commonly eat the seeds of wild grapes and the fruit of the hawthorn and wild rose. It is often said that they consume substantial amounts of dead fish and roe, which has given rise to the belief that their flesh is generally not palatable. However, the Mandarin's primary diet of acorns and seeds, when they are available, undoubtedly makes it as fine a table bird as its cousin, the Wood Duck.

The golden Mandarin ducklings, in their first weeks, rely almost exclusively on the high protein provided by small amphibians, insects, and other invertebrates, switching only gradually to a more varied diet including seeds and other vegetable matter, which they pick from the water or overhead grasses as they swim along the fringes of a stream. They are not yet large enough or strong enough to gulp an entire acorn, but by midautumn they, too, will be digesting the smaller acorns in their gizzards. An adequate food supply is one of the key components in determining the Mandarin's habitat, and at no time is it more important than when the birds are fattening for migration.

Very young ducklings, still in the down, nibble small insects and invertebrates among the weeds. Soon they will begin to eat seeds and aquatic plants.

FALL MIGRATION By early September, Mandarin in Ussuriland are becoming restless as the first snowstorms approach. An unfamiliar chill invades the night air and a morning mist covers the streams. The time to head southward is at hand. All but a few late-nesting females have their new plumage and are ready for the flight. Ducklings are not fully on the wing until eight to ten weeks after hatching, so some, hatched as late as July, are unable to make the flight south until later in autumn.

In preparation for migration, juvenile and adult birds flock together, and when it is time to leave, they depart at dusk and fly well into the night. Although nothing has been written describing the actual movements of Mandarin leaving Ussuriland on their journey south, if they follow the patterns of the North American Wood Duck, they may fly from only a few

The drakes' loss of their brilliant sail
feathers signals the beginning of
their molt in late May or early June.

miles to several hundred miles before stopping to rest. Birds flock for several days on sheltered lakes along their route before arriving at their warmer wintering grounds. Once there, the birds move constantly about, looking for familiar roosts, always with younger birds following the adults.

The time of their autumn passage from Ussuriland varies and continues sporadically until the middle of October. After late October, however, no Mandarin will be seen in this land that is already being covered by ice and snow.

Although the precise routes and destinations of the Mandarin during the autumn passage are uncertain, the age-old wintering grounds in China and Japan clearly beckon. Some ducks, we can assume, find their way down the many valleys west of the Sikhote-Alin range and work their way over or through the inland mountain passes into Manchuria and beyond. Others head south nearer the seacoast to where the Changbai Mountains slope eastward toward North Korea. Here they join Mandarin that have nested in the well-known northern Chinese reserves of the Zhalong Marshes and Changbai Mountains and in the Koreas. Still others veer farther east across the Sea of Japan to winter in its central and southerly islands.

POPULATION More important than the Mandarin's actual migration routes is the troubling fact that birds going north in the spring, or south in the autumn, seem to number far fewer than in decades past. Several Russian experts on the Mandarin, having studied them throughout Ussuriland, have commented on population trends there. Dr. Yuri Shibnev estimates that from 1970 to 1981 the population decreased by fifty percent. Dr. V. I. Labzyuk, however, observes that, although population information is scanty, on the Avvakumovka River in southeast Ussuriland the population has "remained constant for the past twenty years." Professor Yuri Isakov feels that, from fragmentary reports, the Mandarin population in Russian Ussuriland is not less than eight thousand pairs. He notes that parts of southern Ussuriland appear to have lost half of their Mandarin in the decade from 1964 to 1974. In the 1990 *Birds of Ussuriland*, Knystautas and Shibnev write that the number of these beautiful birds in Ussuriland

A typical gang of ardent bachelor drakes converges on a lone female.

decreases constantly. Another estimate appeared in a letter from Dr. Vladimir Borcharnikov who, in 1991, calculated the population of Mandarin in Russia at not more than about three thousand birds. The same year, in a communication with Barry Hughes, he estimated the population at ten to fifteen thousand birds, and then in 1993 in a published report, he wrote, "According to count data from the hunting administration, *A. galericulata* [Mandarin] numbers totalled between 20,500 and 26,800."

Although such population estimates are few, fragmentary, and occasionally contradictory, the majority of observers seem to concur in their belief that Mandarin numbers in Ussuriland and other parts of the Asian flyway are decreasing. Certainly reports of ever-increasing habitat destruction would tend to confirm that belief. Nevertheless, Mandarin are still sufficiently numerous to be sighted as they move south to join other unknown quantities of birds in the lands below and beyond Ussuriland. Without knowing how many birds there are, or exactly where they go, or where they will stop to winter, there is no question that as they leave Ussuriland, the Mandarin spread in a barely visible fan, flying at night, searching for warmer places in a migratory ritual as old as the species.

The Mandarin's powerful wings enable it to lift rapidly off the water. Like the Wood Duck, the Mandarin has spectacularly large wings in proportion to its size.

Migrating Mandarin move in small flocks—flying at night and often for only short distances—through the mountain passes into northern China.

CHINA

For thousands of years each fall or early winter, Mandarin migrating south from Ussuriland into the mountainous Manchurian region of northeastern China have left behind their Russian name, Mandarinka, and assumed the ancient Chinese name of Yuen Yang. The recorded history of the Mandarin in China goes back to the time of Confucius, as early as the fifth century B.C. In those more pristine days, the lake-land plains of China were undoubtedly home to most of the Mandarin in the world. Today in China, however, as entire areas of lakes and forests have been rendered useless for waterfowl of all types, the Mandarin's nesting and wintering area extends only intermittently the length of the eastern seacoast. It starts at China's most northerly boundary with Russia, then descends past the Sea of Japan, the Yellow Sea, and the East China Sea, occasionally approaching the South China Sea near Hong Kong and reaching inland as far as the border with Myanmar and India. From north to south, the Mandarin's flyway in China stretches more than two thousand miles, compared with the Mandarin's flyway in Japan of less than one thousand miles and in Ussuriland of less than six hundred miles.

The great Zhalong Marshes and the Khingan ranges in the far northeast and Yunnan Province far to the south mark the extremes of the Mandarin's major flyway in China today. Of China's present human population of nearly one and a quarter billion, almost one billion (eighty percent of the country's total) are crowded into the rich agricultural valleys and forested mountains that were once the Mandarin's primary habitat. The density of inhabitants in this area is nearly six times the world's average, just over six hundred per square mile. In the lower reaches of the Yangtze River, which at one time were probably the major wintering and nesting grounds of the Mandarin, there are now an extraordinary twenty-five hundred humans per square mile.

As a result of such density, the forest resources of China have, over decades, been massively depleted for agriculture and industry until they are no longer able to provide the country's timber needs. Farms, towns, roadways, and, finally, industries have inexorably encroached on wildlife habitat and polluted many of China's still-abundant wetlands. This is the area where the Mandarin was once so plentiful—and where today it struggles to survive.

Mandarin in the mountainous areas of northern China and in Russian Ussuriland, facing the first chill weather of early autumn, begin their search for the warmer winter areas that will enable them to rest and recuperate following the raising of their broods and the stress of the summer molt. Soon they will start their migration. The routes that the Mandarin travel in China, except for casual sightings, are mostly a matter of speculation. How and where birds pass through to their wintering grounds, how many of them undertake the journey, and where and how many of the birds remain in spring to nest and breed make for fascinating conjecture. But for migrating Mandarin the choice of routes may be critical. Some will survive—many more will fall to storms at sea and to the hunter's and poacher's guns in the vast inland valleys of China.

Unlike recent reports of Mandarin in Ussuriland, which briefly describe their ecology and lifestyle during the breeding season, reports from China deal almost exclusively with wetland areas where they have been found or are thought to inhabit. The migrations and movements of the species are never addressed directly. However, we can imagine that with the first cool days of fall

Range of the Mandarin in China and the Koreas.

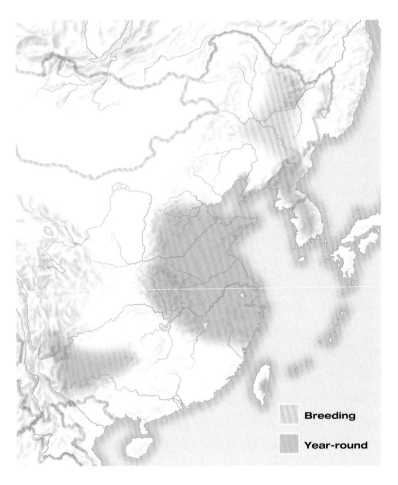

Breeding

Year-round

The relatively long wing of the Mandarin—often with eleven secondary feathers instead of the ten more typical of other ducks—helps the female on the long flight south to China or east to Japan from her northern nesting grounds.

Mandarin from western Ussuriland head south into China through inland mountain passes, or along the valley of the Songhua River in the Manchurian region, to join others in the Zhalong Marshes and Nature Reserve of China's northeastern plains. Here, almost five hundred miles inland, but at an elevation of only four hundred feet above sea level, is China's largest complex of freshwater marshes, shallow lakes, and ponds abounding in frogs, fish, and mollusks, providing food for great numbers of waterfowl. The primary aim of this reserve is to protect the endangered Japanese Crane and other waterbirds by offering safe haven to migrating birds in prime virgin wetlands. Mandarin will stop briefly to join other Mandarin having nested in the neighboring Greater and Lesser Khingan mountain ranges.

While these Mandarin follow an inland route into China, those from eastern Ussuriland may choose a more coastal route, taking them several hundred miles farther south to what is perhaps the best-known nesting area for

Mandarin in China, the Changbai Shan Nature Reserve, just north of the China–North Korea border. This northeasterly area of primeval forests surrounding Tianchi (Heavenly) Lake is a woodland said to be preserved in a natural state, complete with native bird and plant life. The many small bogs and marshes located in the Changbai Mountains are less than a long night's flight for Mandarin from Ussuriland as they join other birds already in the area. The landscape is spectacular, with three major rivers flowing through old-growth forests, high peaks, dramatic waterfalls, and deep crystalline lakes.

These mountains and dense forests offer a haven for more than two hundred sixty species of birds and three hundred species of other wild animals, including the protected Tiger, Sika Deer, Lynx, Leopard, Siberian Musk Deer, Sable, and Eurasian Badger. In addition to providing breeding grounds for the Mandarin, Changbai is an important breeding area for the extremely rare and endangered Chinese, or Scaly-sided, Merganser, whose primary nesting area is the Bikin River basin farther north in Ussuriland.

Mandarin migrating from northern China begin their southerly flight from their spring nesting grounds such as Zhalong, Changbai, and adjacent areas. The long journey south is still before them as they rise at dusk from these magnificent

Far Eastern Frogs spawning in forest pools are easy prey for Mandarin.

and isolated waterfowl reserves to view far beyond them the seemingly endless and crowded valleys of the Yellow and Yangtze Rivers bustling with almost a billion people, to hear the rumble of cars and trucks and the whistling of trains, and to see the vapor trails of jets in the sky. These sights and sounds may cause birds to veer eastward across the Sea of Japan toward Hokkaido in the Japanese archipelago, easily within a night's flight from the mainland. Others may follow the coastline down through the Koreas on their way to southern Japan. There is evidence to suggest that still others, less intimidated by human activity, continue the long and perilous flight over the teeming valleys toward their ancient destinations in the plains and mountainous valleys of southern China.

Until a few years ago, it was thought that few, if any, Mandarin remained in China for the winter. In 1979 Professor Cheng Tso-hsin, one of China's most important ornithologists, wrote me that he and his study group had not been able to find any evidence in recent years of wintering Mandarin in China. Then, in a letter written in October 1990, Professor Cheng noted that Mandarin had been reported breeding in Guizhou and Yunnan provinces. He further suggested I contact Zhao Zheng-jie, "the only person who has been working consistently with the breeding Mandarin in China."

In reply to my inquiry, Zhao wrote that Mandarin breed mostly in the Changbai and Lesser Khingan mountains. Twenty years ago he had seen a flock of fifty birds far to the north, in Changbai, where they bred in the same habitat as the Scaly-sided Merganser, and where he had found nests of both species. He described the wintering habitat to be mostly in the central provinces of Jiangsu, Hubei, Hunan, Sichuan, and Fujian. However, both wintering and breeding Mandarin have been found more recently in Guizhou Province in the highlands of southern China, with its forested mountains, sparkling lakes, and clear streams. The various oaks and beeches in this subtropical climate not only provide nesting sites, but also offer an abundant supply of acorns and other mast essential to the Mandarin's diet.

Mandarin that live year-round in China form pairs by early spring.

Breeding Mandarin are also found in the three large nature reserves in nearby Yunnan Province, at China's border with Myanmar, Laos, and Vietnam. Although they are known to breed in Yunnan, the presence of wintering Mandarin there has not been verified. However, some of the breeding Mandarin in the area undoubtedly will remain for the winter, as they do in Guizhou. The temperate climate of these southerly areas and the abundance of acorns and beechnuts provide ideal habitat for breeding and wintering Mandarin, despite the widespread human activities in many parts of both provinces. According to Professor Won Pyong-Oh, Mandarin are widely dispersed throughout the Korean peninsula. Some are certainly resident; others are passing to or from their wintering grounds in southern Japan.

The presence of Mandarin in Yunnan, so far to the south and west, opens a challenging prospect for further investigation. Over a period of many years, Mandarin have been sighted in small numbers in neighboring Myanmar, even in the state of Assam in India, but were believed to be there by accident, having been lost or blown in by storms. Now it seems possible that in the forested lands of Myanmar, Assam, and other areas even farther west, pioneering Mandarin may exist in small but permanent colonies. Just as North American Wood Ducks have been able to extend their range to new and marginal habitat in North America, so may the Mandarin in East Asia. As with much concerning the Mandarin, these isolated and often inaccessible areas need far more study, particularly with regard to habitat and its suitability.

In December of 1990, Dr. Jianjian Lu, who has been actively involved in studying and cataloging Chinese wetlands, wrote me of additional areas in the south of China where he has discovered wintering Mandarin:

> In China, *Aix galericulata* [Mandarin] is a forest bird, and a bird of small mountain streams, too. According to my records in the last several years, the largest population of wintering Mandarin Duck is in mountain (hilly) areas of southeast China. Typical examples are Pingnan—about 1,200 individuals are found every winter; Wuyuan—about 2,000 individuals; Wuyanlin—about 900 individuals. . . . Total number of this species in China is not less than 20,000.

The nature reserves in China may provide essential habitat for Mandarin and other waterfowl species in areas where forests have been depleted and fragmented.

An alert Mandarin drake displays his gaudy crest.

Whereas Wuyuan and Wuyanlin are several hundred miles inland from the East China Sea, Pingnan is near the seacoast, almost directly opposite the northern tip of Taiwan. These towns are the most southerly locations of Mandarin reported in this eastern area. Dr. Lu writes only of wintering Mandarin, but just as they remain in Guizhou to nest in their wintering grounds, so they will almost certainly nest in these southern coastal areas.

Thus, as the forested wetlands of China's Manchurian region in the far north contain well-established nesting areas for Mandarin, fifteen hundred miles to the south and southwest, in the remarkably mild climate of China's subtropical mountains and marshes, are other well-established wintering and probably spring nesting grounds.

Situated between these northern and southern areas of Mandarin habitat are the great long valleys of the Yellow and Yangtze Rivers, which contain some of the largest wetlands in the country and another concentration of freshwater lakes. With the vast river wetlands having been destroyed, the flight of the Mandarin from north to south is simply a rite of passage. In contrast, less than one hundred years ago, Mandarin were once so numerous on the Yangtze River that they could be caught by hand. Fowlers floated empty calabashes on the water until the birds became used to their presence. The fowlers then submerged themselves in the water, heads covered with the hollowed calabashes pierced with breathing

holes. Moving slowly among the Mandarin feeding in the reeds along the shoreline, they pulled the birds under the water. Many Mandarin caught this way were purchased by the gunboat crews of the Royal British Navy stationed on the river.

Today, it would be unusual to sight even a single Mandarin on the Yangtze. The fertile wetland areas have been stripped of most of their characteristic ancient deciduous, broad-leaved trees to accommodate a massive farming and industrial complex. Still, the many remaining lakes, marshes, and streams of the plains are a magnet and a trap, drawing migrating waterfowl for food and rest—but often for slaughter. Six adjoining provinces in the south-central part of the Yangtze River valley are the major area in China for the production of wild duck meat. In the mid-1950s, over one million wild ducks were harvested in this region alone. Honghu Lake, the largest lake in the area, produced for market nearly a half-million ducks. Nearby Fuwan Village accounted for another hundred thousand wild ducks shot. When wetland areas around Fuwan Village were drastically decreased by drainage to produce more arable land, the harvest of waterfowl declined, only to be offset by an uncontrolled increase in winter shooting, improved hunting techniques, and a threefold increase in the number of guns used by hunters.

The Mandarin in coastal China, once a common sight on the open water and shorelines of large rivers and lakes, are today almost impossible to find.

However, within this heavily farmed, industrialized, and commercially hunted area of nearly one billion people, there are two important nature reserves of potentially great benefit to waterfowl, including the Mandarin: Poyang Lake in Jiangxi Province and Dongting Lakes in Hunan Province, a region in the southern end of the great plains of the Yangtze River basin and some five to seven hundred miles inland and slightly south of Shanghai. Both are affected by the radical changes in water level and surface area that typify other freshwater lakes in the plains.

Poyang Lake, at an elevation of only a few feet above sea level, is China's largest freshwater lake. The lake and its many surrounding marshes and wetlands, fed by five major rivers, cover an area of almost one million acres. The Poyang Lake Nature Reserve, established in 1982, includes within its more than fifty thousand acres sixty small villages with a population of twenty thousand people. The reserve is administered by the Provincial Forestry Department, and although more personnel trained in ecology and ornithology are needed, efforts to moderate the damage to the environment have already achieved promising results. The reserve is unique in the quality of its ecosystem, and waterfowl are effectively protected from hunting, though poaching is still a major problem. Since the discovery of the world's largest winter concentration of Siberian Cranes at Poyang Lake, wildlife tourism has increased dramatically and hunting has been even more strictly controlled. The South China Crane Research Center and Wetland Education Center will soon be established at the reserve and will help to increase public education about wildlife recognition and management.

In the dry winter season, Poyang Lake decreases in surface area from over eighteen hundred square miles to as little as two hundred square miles of smaller shallow lakes rich in small fish, shrimp, and mollusks. Many species of wildfowl winter here, but Mandarin almost certainly stop only to rest and feed before continuing their migration to more heavily wooded areas. The loss of sheltering trees and shrubbery and the pressures of poaching will send them farther south, in search of more favorable winter habitat.

Dongting Lakes, over two hundred miles due west of Poyang, is a group of large lakes and wetlands covering an area of more than one million acres. In the early 1800s, Dongting Lake, the largest in the complex, was China's largest freshwater lake, with a surface area of over two thousand square miles. Today it is estimated to have a surface area of less than fifteen hundred square miles, and according to some predictions, it will disappear in the next fifty years unless strong protective measures are taken. In 1984 approximately one-third of the area was designated as a nature reserve under the control of the municipal government. Hunting regulations are in effect but are difficult

OPPOSITE A showy Mandarin drake searches a stream for acorns or small amphibians.

to enforce, and poaching is still prevalent. In the dry winter season at Dongting, as at Poyang, lake levels may drop by more than fifty feet, creating myriad shallow ponds interlaced with mudflats and sandbars rich in small fish, invertebrates, and other food sources attractive to migrating or wintering waterfowl. Although Mandarin have been reported wintering at Dongting, the exposed shoreline, lacking the cover of sheltering trees, would cause these Mandarin, like those at Poyang, to stop only in passage as they continue their southern migrations to more protected areas.

Poyang and Dongting Nature Reserves are rare beacons in these vast valley environments. Although Mandarin have been sighted in the same general area on more than a dozen other wetland reserves—such as Nansi Lake in Shantung Province, Shuangtaizi and Liao marshes in Liaoning Province, and Shijiu Lake on the border of Kiangsu and Anhwei provinces—these are reserves in name only since hunting controls have not been implemented. The majority of wetlands are not part of any reserve and, likewise, offer little or no protection for wildlife. Without the type of conservation and hunting control established at Poyang and Dongting, the valleys between the mountains in the north and south, studded with unprotected and often polluted lakes and marshes, become almost a death trap for migrating birds, including the officially protected Mandarin. Too few hunters are willing or able to distinguish the Mandarin from legally hunted waterfowl. Thus these great valleys, once the prime habitat of the Mandarin, have become a treacherous migratory corridor—a choke point—threatening its very survival. In these valley areas may lie the test for conservation to protect this threatened species.

Today, China differs little from many of the world's other developing nations—still clearing the most fertile land for agriculture, cutting forests to supply firewood and to make paper products, draining huge wetland areas to produce more arable land, and harvesting more wildlife to feed an ever-increasing population. Meanwhile, air and water pollution continues. The great hope for the Mandarin and other waterfowl on the country's seventy million acres of wetlands and nearly twenty-five hundred freshwater lakes is an apparent reawakening attitude toward conservation. China established the Dinghu Mountain Nature Reserve in 1956, making it the first nature reserve in China. This and other reserves are part of the worldwide network of nature reserves for "Man and Biosphere" coordinated through the

Mandarin that do not winter in China may continue across the East China Sea to the southern islands of Japan.

United Nations. By 1980, there were seventy reserves, and by 1987, there were four hundred sixty-eight encompassing almost one hundred thousand square miles of China. Within some of them, Mandarin have found welcome refuge.

The future of the Mandarin in China is largely a problem of habitat and will almost assuredly be determined by the effectiveness of the expansion and management of this already extensive system of reserves to resist the encroachment of human settlements, hunting, agriculture, and industrial pollution. In addition to the promising proliferation of wildlife reserves, there is also encouraging activity by many dedicated and qualified individuals who are learning and writing about conservation and, in the process, are preserving the Mandarin in China. Although the task of conserving China's natural resources is daunting, such activity offers the only hope for the future of the magnificent Yuen Yang, so highly prized by the Chinese of earlier times and so endangered by the seemingly unstoppable rate of habitat destruction typical of the emerging economies of the late twentieth century.

In Japan, where Mandarin have been revered in art, literature, and religion for centuries, they still bring beauty to the countryside and pleasure to everyone who sees them.

JAPAN

With the onset of fall, the williwaw winds sweep down the Aleutian Islands, along the giant Kamchatka Peninsula, and over the Kuril Islands. There they join the first storms moving south from the tundra of eastern Siberia, across the Sea of Okhotsk, and running the length of Sakhalin Island. The Mandarin that nested and raised their broods in Russian Ussuriland, Sakhalin Island, and Kunashir in the Kurils look south to Japan and specifically to Hokkaido, the second largest and most northerly of Japan's major islands. For many millions of ducks, geese, and swans from this part of the world, Hokkaido is a major staging ground for their southerly migration. By October, when the rivers, streams, and lakes of this rugged island begin to freeze, the Mandarin that nested here have been joined by birds from the other areas of the Mandarin's northern range and are ready to fly south to the islands of Honshu, Shikoku, Kyushu, and beyond.

From Hokkaido in the north to Kyushu in the south, Japan stretches approximately twelve hundred miles and averages less than two hundred miles in width. The topography of the island chain is defined by massive mountains, many of volcanic origin, whose peaks rise over ten thousand feet. Running the full length of the islands, they constitute nearly eighty percent of Japan's landmass, leaving little flat or arable land along the shoreline. In the lowlands, which are inhabited by humans in greater density than any other area of the world, nearly every acre of arable land has been put to agricultural or industrial use, and most major rivers are revetted or dry. Still, waterfowl come to winter in the more than six thousand wetlands of Japan—its streams, ponds, lakes, reservoirs, and rice fields. Today, more Mandarin are believed to winter in Japan than in all other areas of its range, including Russia, China, Korea, Okinawa, Taiwan, and other, less frequented locations.

Mandarin migrating south to Japan from their most northerly breeding areas on the islands off the Asian mainland follow two ancient routes of passage: Mandarin from the southernmost Kuril Islands are joined by other species of migrating ducks from the Kamchatka Peninsula. Birds taking this route may have originated in eastern Siberia and even Alaska. For the Mandarin in the southern Kurils, Hokkaido is less than an hour's flight away. Mandarin having nested on the southern tip of Sakhalin Island may be joined by others departing their breeding grounds in Ussuriland. From Sakhalin they, too, fly to Hokkaido, across the narrow strip of La Pérouse Strait, in slightly more than an hour.

Mandarin from the Asian mainland fly to Japan along two other well-traveled corridors. Those from central and southern Ussuriland, even from northern and central China and the Koreas, make the long flight across the Sea of Japan. Flocks of Mandarin, appearing to number in the thousands, have been observed resting on open water in the sea, apparently on their route from the mainland. Since Mandarin typically fly at little more than thirty miles an hour and fly almost exclusively at night, they normally could not reach Japan without stopping at sea to rest and await the coming of dusk. Birds from South Korea make the shorter flight across the Korea Strait and have been found resting on Kyushu and southern Honshu and on the tiny Japanese island of Tsushima midway in their passage.

More than thirty species of waterfowl, some flying thousands of miles from their spring nesting grounds, congregate in Japan to spend the winter. Along the shores of lakes and reservoirs, on rivers and streams, in mountain and coastal areas, in ponds and parklands, even in the gardens of the Imperial Palace, Mandarin find shelter under cover of the deciduous, broad-leaved trees they cherish. Several million waterfowl are believed to winter in Japan and its thousands of wetlands. Following an often long migration, such a large number of waterfowl require a substantial food base. Rice, Japan's major agricultural product, is harvested in October, just prior to the arrival of most

Breeding

Year-round

Range of the Mandarin in Japan.

Mandarin hide in the shadows of a temple pond.

migrating waterfowl, and the next planting is not until April, after most have left for their northern nesting grounds. Hence, the wintering birds can forage on kernels left over from the harvest. A 1949 report, "Waterfowl of Japan," prepared by the Supreme Command for the Allied Powers, noted, "The most careful harvester, however, is unable to take every kernel of rice from his paddy. Some always drops in the mud, where no tool known to man can retrieve it but where it is ideally located for the probing flat bills of the dipping ducks. Enough such waste rice is left in the paddies each autumn to maintain many times the present duck population, which, in turn, is the best means of salvaging this wastage and turning it into easily available proteins."

In Hakone, a mountain resort less than one hundred miles southwest of Tokyo and only twenty-three miles from Mount Fuji, Yuzo Murofushi has been studying, trapping, and banding Mandarin since the early 1980s. Murofushi, one of the few naturalists to have studied the Mandarin in Japan, also has photographed them in and around one of their favorite wintering and nesting areas at nearby Lake Ashi. Despite more than three hundred bandings, only one recovery has ever been reported. This is largely because hunting of Oshidori, as the Mandarin is called in Japan, is prohibited by law. Birds that are legally shot or those that are involved in widespread trapping programs are the most common sources of band recoveries, which are used to arrive at valuable information regarding population trends and migration routes. Unfortunately, such information is not available on the Mandarin.

In the winter of 1987, Lawton Shurtleff and his wife, Anneke, came from California to Japan hoping to see, in their native habitat, the Mandarin that I had been writing about. One chilly winter morning, Murofushi met the three of us at the charming old Fujiya Hotel in the mountains near Hakone. From the hotel he led

us on foot to a snow-covered bridge nearby, which spans a one-hundred-foot-deep gorge, cut by Jakutso, a tributary of the Hayalawa River that flows out of Lake Ashi less than three miles to the north. Billows of steam rose from the hot springs bubbling in the streambed. Barely visible at that distance through the mist were a dozen or more Mandarin perched under the branches lining the rocky shoreline. At short intervals, one or more would dart

into the turbulent stream, dip under the water, and quickly return to shore. Murofushi explained that the steam attracted insects, which the ducks were intent on devouring. Such insects were a special source of protein when snow covered the landscape.

This precipitous gorge was a favorite spot for Murofushi and his high

ABOVE More than thirty species, totaling several million waterfowl, winter in Japan. Among them are as many as sixteen thousand Mandarin. **BELOW** As ponds in northern Japan freeze over, Mandarin search for open water on large reservoirs.

ABOVE Unknown numbers of
Mandarin remain in the spring to
nest and raise their broods in
Japan.

BELOW Some can be been found
on ponds in lushly planted, small
private gardens.

school students to study the Mandarin. They reached the stream by descending ropes secured to the trees. To catch Mandarin for banding, they used the traditional Japanese technique of flushing the birds upstream into nets they had stretched across the narrowest section of the ravine. Trapping, banding, and releasing wild birds are essential research techniques for determining migrational patterns and other aspects of their behavior. Despite the importance of this well-recognized methodology, Murofushi and his students are among the few to attempt to band the Mandarin in Japan.

To learn more about the Mandarin in Japan in winter, the three of us visited a famed sake producer in Hida Takayama in Gifu Prefecture, two hundred miles west of Tokyo, just north of the isthmus joining southern and northern Honshu. The owner, also a student of Mandarin, introduced by our interpreter as Mr. Oida, showed us a small museum in his office devoted to the Mandarin and a large sacred oak tree standing nearby. Each spring the tree housed a nesting Mandarin hen. From this nest the hen would lead her newly hatched brood to a nearby temple pond. To escape with her young ducklings from the busy pond to the nearby large river was almost a feat of magic. In front of the dozens of modest shops between the pond and the river was a small drainage canal carrying water from the pond and barely large enough to hold a grown Mandarin. Each morning the shopkeepers washed the narrow alleyway in front of their shops by throwing out buckets of fresh pond water that filtered into the narrow canal. Into this channel partially covered with stone slabs, the Mandarin led her ducklings to float literally under the feet of the busy shopkeepers above them. In the nearby river, we could see dozens of Mandarin and other waterfowl among the snow-tipped boulders. On the riverbank, side by side, stood open-air stalls of all types, teeming with visitors. Yet the usually reclusive ducks seemed at ease among the human commotion. Quite unlike other areas of the Mandarin's range, in densely populated Japan one can readily see and almost touch the Mandarin, at least during those five

One of Japan's most important contributions to the study of the Mandarin has been the midwinter tabulation of waterfowl in more than four thousand wetland sites by some fifteen thousand counters. This chart shows the population trends in the forty-seven prefectures over the past twenty-three years.

MANDARIN WINTERING IN JAPAN
Midwinter Counts 1973–1995 by Prefectures

Prefecture	No.	1973	1974	1975	1976	1977	1978	1979	1980	1981	1982	1983	1984	1985	1986	1987	1988	1989	1990	1991	1992	1993	1994	1995
HOKKAIDO	1	23	45	64	87	90	198	483	95	32	15	32	37	47	40	97	57	42	68	85	124	19	170	32
AOUORI	2	650		6	90	457	192	162	4	24	6	19	7	2			7	9	1	17	6	2		5
IWATE	3	688	830	290	520	419	107	177	338	315	138	171	135	105	109	177	92	110	91	127	133	115	55	204
MIYAGI	4	17	90	27	225	48	60	48	22	36	70	34		5			8	25	4	20	21			6
AKITA	5	15	32	126	24	74	44	131	147	85	25	46	81	28	37	36	36	33	6	28	65	36	20	31
YAMAGATA	6	148	97	148	107	98	109	33	228	141	58	140	75	85	31	52	57	65	143	97	134	196	80	197
FUKUSHIMA	7	525	51		388	142	106	82	58	45	217	38	202	134	109	178	376	401	275	398	363	343	171	109
IBARAKI	8		2	48	9	31	9	33	75	5	35	99	80	77	63	62	41	5	57	79	462	179	393	612
TOCHIGI	9			10					8			3			17	1	43	13	11	150	581	260	68	714
GUAMA	10	614	546	850	987	407	181	398	389	711	466	174	155	368	367	519	982	219	383	246	1,363	242	577	586
SAITAMA	11	165	107	180	140	90	83	100	67	42	87	38	32	50	63	162	58	138	137	114	1,050	54	513	345
CHIBA	12	27	3	69	28	90	79	513	124	195	317	122	339	99	110	472	428	830	191	468	294	475	255	350
TOKYO	13	135	75	104	90	148	111	198	294	310	280	420	299	358	978	552	443		602	396	724	485	425	216
KANAGAWA	14	579	543	325	278	807	930	290	628	796	262	337	169	194	414	346	782	334	176	547	1,629	356	906	670
NIIGATA	15										6	8	2	18			3	675	1	7	3	5	4	6
TOYAMA	16	34	14	62	129	53	190	265	254	188	179	243	165	98	114	39	131	5	113	156	178	80	91	135
ISHIKAWA	17		20	30		100	545	20	400	60		4		5		15		74	4	1				
FUKUI	18	50	51	51	84	33	40	110	65	64	18	15	84	28	45	37	17	2	2	2	14	11	6	15
YAMANASHI	19	6	210	367	37	45	140		81	45	41	68	21		136	173	22	10	57	84	131	51	139	115
NAGANO	20	82	42	155	53	22	44	106	172	210	273	338	708	133	424	141	353	53	222	262	432	507	206	314
GIHA	21	207	197	188	271	334	252	353	229	144	24	186	108	253	254	159	245	1,454	318	585	446	641	422	217
SHIZOUKA	22	57	5303	738	689	332	139	1207	118	152	231	105	227	407	636	495	653	332	159	585	135	141	308	521
AICHI	23	81	149	149	140	145	156	174	192	135	135	79	242	146	393	99	76	482	100	345	289	418	314	389
MIA	24	95	128	240	388	225	153	223	505	295	97	146	134	309	42	62	9	181	31	122	59	399	166	201
SHIGA	25	260		12	28	36		6	60			11	26	2	19	28	2	37	207	93	248	92	211	80
KYOTO	26	104	261	202	130	274	114	181	205	159	166	359	270	265	282		138	60	102	136	194	214	180	203
OSAKA	27	50	162	83	104	98	188	188	109	65	75	165	226	415	548	291	218	225	341	563	812	579	482	614
ILUGO	28	8	71	28	45	83	68	100	76	38	50	48	84	85	64	207	124	441	166	65	127	58	88	110
NARA	29		300	67	128	298	278	282	637	392	310	452	602	412	338	1,186	470	72	886	494	567	249	1,058	1,205
WAKAYAMA	30	599	308	566	1,146	783	732	460	407	323	351	664	461	731	1,024	855	286	1,058	428	447	957	755	1,154	1,445
TOTORI	31		1						2				16	24		15		1,178	27	16	14	7	25	139
SHIMANO	32															1	1	20	252	180	275	343	554	376
OKAYAMA	33	125	123	57	80	6			6	97	123	147	215	24	135	49		746	87	128	99	166	63	39
HIROSHIMA	34	178	54	55	47	55	49	7	145	11	21	31	80	52	174	29	178	140	79	253	28	148	142	221
YAMAGUCHI	35	120	99	144	402	390	214	250	327	538	27	229	383	688	1,133	916	520	143	608	1,156	435	645	770	481
TOKUSHIMA	36	17	210	197	167	52	70	78	109	184	258	46	171	192	76	171	201	1,058	120	425	305	477	539	637
KAGAWA	37	5	4	7	8	11	18	31	110	94	18	71	81	73	127	89	143	7	40	20	113	103	153	19
EHIME	38	377	301	1,809	2,151	792	996	3,009	301	272	285	240	451	459	406	719	780	38	1,783	1,617	2,576	1,333	2,780	1,461
KOCHI	39	172	182	321	494	475	346	149	353	453	433	241	348	437	389	426	824	1,082	684	305	2,070	959	1,472	1,641
FUKUOKA	40	20	8	20	30	39	5	75	22	52	22	16	15	26	46	118	79	817	253	166	199	149	298	201
SAGA	41	35	180		88	35	396	260	179	284	87	166	198	255	363	237	330	96	657	187	351	347	140	308
NAGASAKI	42	82	120	20	86			26	48		67	1,246	930	1,983	921	1,685	968	218	1,733	3,656	753	1,455	1,047	1,395
KUMAMOTO	43	565	123	462	484	379	152	210	140	341	143	235	165	450	493	421	742	1,843	297	499	595	587	341	220
OITA	44	57	93	182	192	259	217	130	216	516	326	380	639	479	511	708	288	689	403	965	937	684	728	566
MIYAZAKI	45	254	470	368	227	461	482	207	207	991	214	270	379	270	2,956	582	152	1,158	203	526	506	502	840	796
KAGOSHIMA	46	3,384	1,824	2,851	646	834	691	404	603	630	1,035	348	649	528	401	714	732	426	826	245	594	703	461	544
OKINAWA	47								13		2								446	6				12
TOTAL*		10,053	13,422	11,682	11,277	9,550	8,803	11,167	8,754	9,379	6,970	8,219	9,631	10,946	14,692	13,377	12,181	16,163	13,361	17,074	21,390	15,591	18,815	18,703
EA INDEX (1988=100)		82	110	96	93	78	72	92	72	77	57	67	79	90	120	110	100	133	110	140	176	125	154	154

(Left margin graph axis labels: 200%, 150%, 100%, 50%, 0%; EA Index 1988=100)

n.b. not all columns total, i.e., 1978 = 1984   Source: Environmental Agency of Japan

or six months, from October into early March, before the beginning of the spring migration.

Japan's greatest contribution to our knowledge of the Mandarin throughout the Far East is an unprecedented program of counting populations of all species of waterfowl including Mandarin. Organized in 1972 by the Environmental Agency of Japan—working in cooperation with a private organization, the Wild Bird Society of Japan (WBSJ)—over fifteen thousand spotters, many of them volunteers, fan out to more than six thousand wetlands in mid-January to identify, count, and verify Japan's immense winter waterfowl population. In 1992, among the many species sighted and counted were 21,390 Mandarin. Of this number, approximately 17,100 were on man-made lakes (hydroelectric reservoirs), 3,000 on the upper ends of rivers and streams, and 1,300 on natural lakes and marshes. Although reservoirs are not typical Mandarin habitat, their stable water levels support the lush woodland growth that provides excellent shelter.

Mandarin traverse the many narrow canals in Japan's farmlands and find plentiful food supplies in the rice paddies.

All visual counts of Mandarin admittedly are inconclusive, given the secretive nature of the species and its reluctance to venture into open waters. Nevertheless, these tallies are the only hard data available on which to base estimates and trends of the total population of Mandarin in the Far East.

The average Mandarin count for the twenty-one Januaries from 1975 through 1995 is approximately 13,000 birds. The greatest number counted was 21,390 in 1992 and the lowest was 6,872 in 1982. A significant finding is the apparent upward trend from the low in 1982 to the high a decade later, when the Mandarin count had tripled. Wildly varying counts in the late 1970s and early 1980s caused some experts to be skeptical of their accuracy. Consequently, from 1982 to 1992, the Wild Bird Society of Japan conducted independent surveys on the same day in mid-January. Although the WBSJ covered only one thousand to fifteen hundred sites, compared with the over four thousand sites included in the Environmental Agency of Japan count, the WBSJ trend lines were so similar that in 1992 the WBSJ discontinued its separate program and accepted the results of the agency while continuing to cooperate with it.

Mandarin, like Wood Ducks, typically avoid open water. Yet eighty per-

cent of the wetlands where Mandarin were counted were man-made lakes. I have participated in several such annual winter waterfowl counts on Lake Ashi. Ashi is a large volcanic lake in Hakone Prefecture covering eighteen thousand acres at an altitude of over two thousand feet. It is surrounded by thick forests of naturally occurring deciduous trees. Each year, our count was conducted by a group of fifty to sixty spotters traveling by motor launch, who toured the entire shoreline in three to four hours. Mandarin, hiding along the shoreline under the branches, were extremely difficult to identify. Although we spotted over thirty Mandarin, all of us agreed that we may have seen only one-tenth of those actually in residence since they hide in small groups under the overhanging branches, and their keen eyesight detects the slightest hint of danger, which causes them to vanish, often uncounted, into the nearest shade and shelter.

So what do the twenty-odd years of counting Mandarin tell us about the winter population in Japan? At Lake Ashi, we felt we were seeing only one in ten. In Great Britain, repeated visual counts of Mandarin by Andy Davies recorded only twenty-five percent of the birds known to be on certain relatively small ponds. It is obvious that visual counts of Mandarin may miss many more birds than they find.

If the average wintering population for the past twenty-three years approximated thirteen thousand birds and if the trend indicates a present population of sixteen thousand birds, a multiplier of four would suggest a population of fifty thousand to sixty-five thousand. Although any estimate is

TOP LEFT AND RIGHT The Yellow Weasel is a ferocious predator attracted to nestboxes with overly large entry holes.

ABOVE The Raccoon Dog takes its share of Mandarin.

by definition a guess, it is clear that Japan has a significant population of wintering Mandarin. This raises an intriguing question: Where do such large numbers of Mandarin nest and breed in springtime? The varying and essentially fragmentary reports from eastern Russia and China offer scarcely a clue.

The answer might lie in Japan itself. In 1984, two of Japan's leading authorities on the Mandarin, Professor Yuzo Fujimaki and Yuzo Murofushi, collaborated on a paper entitled *Distribution and Abundance of the Mandarin Duck in Japan*. The presence of breeding Mandarin was confirmed in Hokkaido in seven areas along mountain streams and on reservoirs. Sixteen more areas in Hokkaido were believed to accommodate nesting Mandarin, but none could be confirmed. In another half-dozen areas, birds were sighted during the breeding season, but no nests or broods of ducklings were seen. Farther south on the main island of Honshu, in Hakone, south of Tokyo, nesting Mandarin were confirmed on Lake Ashi and along nearby Sakawa River. All sightings of broods of young ducklings, observed from a motor launch, were along the sheltering shoreline. Here, as in Hokkaido, relatively little activity was observed. On Lake Ashi, for instance, in the six years since 1978, the ducklings counted increased from five to only twenty-one.

In addition to describing their own limited observations, the authors refer to a survey of breeding birds conducted in 1978 by the Wild Bird Society of Japan and to information from the Environmental Agency of Japan covering the years from 1974 to 1978. In this survey, throughout Japan, breeding was confirmed in only eighteen areas, mainly in Hokkaido and northern Honshu, and was thought probable but not confirmed in fifty-nine other areas. Despite their efforts, the authors were forced to conclude that "there is little information on the abundance of the Mandarin Duck in the breeding season."

A partial explanation for this apparent lack of information is suggested by Mark Brazil, an English biologist and recognized authority on wildlife in Asia. Brazil is particularly struck by what he refers to as "the paradox of the Japanese Mandarin," for while Japan provides perhaps the perfect home for the Mandarin, it is the symbolic Mandarin, not the living Mandarin, with which the public is familiar. The bird itself, he feels, has gone "not so much

A drake and his mate add their beauty to the tapestry of spring color on the blossom-covered pond.

unnoticed as unwatched and unstudied." The Mandarin is a "forest lover," and it is in the forested mountains where it is truly at home and probably at its most common. These forests support a wide range of acorn- and nut-bearing trees that provide the food Mandarin prefer. The north-to-south extent of Japan and its various climatic zones have enabled the evolution of four main forest types: subarctic coniferous, cool-temperate broad-leaved deciduous, warm-temperate evergreen, and subtropical. In all but the first of these, Brazil has found Mandarin during more than a decade of study. He concludes that, in theory at least, much of Japan offers suitable Mandarin habitat.

A nesting hen feeds early in the morning and again at sunset. She carefully searches the area before leaving the nest cavity.

Having explained the theory of Japan's forest potential, he proceeds to explain why reality may be difficult to reconcile with theory. One special feature of Japanese forests is the undergrowth, consisting of various species of dwarf bamboo of tropical origin, found in all forest types but especially in the cool-temperate forests of northern and central Honshu. Dwarf bamboo inhibits natural regeneration and, in combination with the steep slopes, makes hiking and birding in the forest very difficult except on established trails. This dense ground cover of wiry, shoulder-high bamboo grass, known as Sassa, seriously limits access and can make searching for wildlife exceedingly difficult. The valleys and forests that are impossible for the ornithologist to access can be easily flown through by the agile Mandarin. The terrain—with its tortuous valleys and streams—may provide the Mandarin with ample riverside breeding habitat. It is unfortunate for the Mandarin that so little is known of the forests in these rugged areas. During World War II, most accessible forests were indiscriminately logged to feed the engines of war and have since been replanted with conifers and other trees easily harvested but lacking suitable nest cavities for Mandarin. Official estimates are that sixty-two percent of forests today are still in their virgin state. Others believe that the old-growth trees constitute less than twenty-five percent. In any case, deep in the mountains and valleys where foresters and ornithologists are unable to penetrate, Mandarin may still find old-growth trees, and so the theory persists that here Mandarin may nest and brood, hidden from view, in numbers far greater than reported. But the possibility has not been confirmed.

Predators of the Mandarin in Japan are relatively few. They include a number of large raptors such as the Golden Eagle and Northern Goshawk. The most notable potential predator, in Honshu at least, is the Hodgson's, or Mountain, Hawk Eagle. These raptors take birds and mammals as large as a hare, and thus the Mandarin is a potential meal. The largest owl occurring throughout most of the Mandarin's Japanese range is the Ural Owl, which may not be large enough to handle an adult Mandarin but could readily capture chicks and fledglings.

Mammalian predators include the omnivorous Raccoon Dog, or Tanuki, and the Red Fox, especially as Mandarin are most vulnerable while nesting and raising their ducklings. Martens and related species, nimble tree climbers capable of taking prey up to the size of a rabbit, probably pose the greatest threat to adults on the nest and to eggs and chicks. A number of snakes capable of climbing trees also may pose a threat to the eggs. Feral cats and large fish are believed to take a heavy toll of Mandarin hens with their broods. It is Mark Brazil's opinion, however, that such predation is unlikely to have a major impact on the Mandarin population.

Despite the determined efforts of a few experts, too little is known about the nesting and breeding population of the Mandarin throughout its entire range. However, it should be emphasized that several essential ingredients for estimating the population dynamics of a bird as shy and secretive as the Mandarin are missing throughout the Far East. Legalized hunting of ducks in North America provides the kind of information not available on the Mandarin in its protected status throughout its entire range. The U.S. Fish and Wildlife Service, in cooperation with agencies in Canada and Mexico, estimates harvest (kill) data from hunters and makes indirect population estimates by species. These estimates are the result of an analysis of duck stamps purchased by hunters, a detailed questionnaire to selected hunters, leg banding and band recovery data, and counts of wings sent in by cooperating hunters. Such indirect population estimates are particularly valuable for the Wood Duck, which, like the Mandarin, is not susceptible, by nature or preferred habitat, to the direct counting methods used successfully for other species. This is not to recommend that the Mandarin be subjected to hunting, but to point out that hunting can provide valuable information not presently available on the species. In addition, widespread use of nestboxes for Wood Ducks by private and public agencies suggests data on population trends across North America and helps determine migration patterns by offering an opportunity to band nesting hens and web-tag ducklings and to capture nesting hens banded in other areas.

The shy and secretive Mandarin, like the North American Wood Duck, is difficult to census by direct counting. Yet the Japanese winter counts, given the number of wetlands observed and the participation of so many observers, is an extraordinary effort that must form the basis for future population estimates. There is much left to discover about the Mandarin in Japan. A national nestbox program could help establish whether there is a shortage of natural nest cavities throughout the islands. Banding of nesting or trapped birds, in cooperation with the Koreas, China, and Russia, could, in time, yield solid information on migrational patterns and indirectly, perhaps, on population trends. The paradox of the Mandarin in Japan will be more readily resolved when society knows and admires the living Mandarin with the same devotion and intensity it has worshipped the symbolic Mandarin of the past.

LEFT A drake in full sail engages in ritualistic courtship display.
ABOVE, FROM TOP He dips his bill, jerks it backward, then preens behind one wing.

The thriving population of introduced and naturalized Mandarin in Great Britain came from the release of fewer than fifty birds from the wilds of China in the early 1930s.

GREAT BRITAIN

No description of the Mandarin's world would be complete without discussing its presence in Great Britain—for while the Mandarin is not native to this part of the world, perhaps as many of them live and breed in Britain as in any region on the Mandarin's native flyway in the Far East. In addition to the thousands of Mandarin flying wild throughout Britain, there are—and for many years have been—Mandarin kept in private gardens and estates. The first pair were brought to Britain in 1745, and the first Mandarin to breed were at the newly formed Zoological Society of London in 1834. Since then, the British, with their unmatched interest in and concern for wildlife, have embraced the exotic Mandarin as one of their own. In 1971, the Mandarin was officially admitted to the British and Irish List after it had been breeding in the wild in Britain for nearly forty years. It is fair to say that, today, Mandarin are better known and more admired in their adopted home in Britain than in those countries where they occur naturally. The story of the

Mandarin in Britain is as much about the men who brought them there and cared for them as it is about the duck itself—and I am proud to be part of that group.

One of the greatest private collections of waterfowl was created in the early 1900s by the eleventh Duke of Bedford at Woburn Abbey in the county of Bedfordshire, north of London. By the beginning of World War I, Mandarin in his collection were believed to number over three hundred. They lived freely in and around the abbey, where they were regularly fed by the groundskeepers. Their numbers were so great that farmers nearby often complained that their fields of grain were being ravaged by the ducks.

Within farmers' manicured fields Mandarin still find trees with suitable nest cavities and small streams for raising their ducklings.

During World War I, and again, during and after World War II, all artificial feeding was discontinued at the abbey, and the Mandarin population declined to approximately one hundred fifty birds, a level at which it remains today. Woburn Abbey is a forest in miniature, studded with small ponds and streams. The woods yield a rich harvest of acorns, sweet chestnuts, and beech mast. Surrounding the ponds are thickets of rhododendron, where the Mandarin may rest in safety and seclusion. Predators are kept to a minimum so the ducks live in a protected environment. Their continued success at the abbey under fairly natural conditions demonstrates how well they can adapt if they are given a suitable environment. Although local conditions are adequate to sustain this small colony, there is little evidence that they have spread beyond this area.

Another notable collection of waterfowl containing Mandarin was that of Lord Grey of Fallodon. In February 1884, while still a student at Oxford University, he began his bird collection in Northumberland, in northern England, close to the Scottish border, where he provided food and nesting facilities for his birds. It was not until 1918, however, that the era of Mandarin really began for him. The original stock for his collection came from two fanciers, McLean and Wormald, and these Mandarin nested and hatched ducklings every year, both in special nesting barrels and in hollow trees outside his estate. Lord Grey's was a complete and perfect sanctuary. By sheer perseverance and patience, he tamed Mandarin as no one else has ever done. They willingly came to him, ate from his outstretched hand, and perched, seemingly without fear, on his head and shoulders. Given their extreme shyness, this was a remarkable achievement.

Year-round

Range of the Mandarin in Great Britain.

To Lord Grey there was nothing unusual about this. In 1930 he wrote that a Mandarin duck had discovered that flying up and standing on the head or shoulders of a man who had food gave it a position of vantage over the other birds and attracted immediate attention. Other ducks followed this example, and at the evening feed, they often competed for this position. In wet weather, Lord Grey wrote, the birds sometimes left traces of mud on his hat and clothes, but he and his gardener noted that the Mandarin treated the human form with as scrupulous a regard for cleanliness as they did their own nests.

Lord Grey was a keen observer of the waterfowl in what he called his sanctuary. In his *Fallodon Papers*, published in 1926, he describes his several acres of flower gardens, small ponds, trees, and shrubs, surrounded by a fence just high enough to keep out foxes. He tells of pintails, shovelers, and teals that at Fallodon would be so tame they also ate from his hands. Yet those same birds on ponds nearby would immediately fly from him if he attempted to approach them. He writes that "tameness and confidence are associated with the place and do not cause them [the ducks] to be less wild elsewhere." He tells about another duck that had fled from him when it was on a neighbor's pond, but was unafraid and tame when it was on his own ponds. Lord Grey's experience shows how quickly birds learn that a certain place is their sanctuary.

BELOW AND BOTTOM During his twenty-year study of the Mandarin in southern England, Andy Davies and his co-workers banded many birds. One of the ducks was recovered in 1994 in Estonia, more than one thousand miles east of England.

In October of 1930, only three years before his death, despite his failing eyesight, Lord Grey described relations between the Mandarin and the Wood Duck. In comparing the ducklings of the two species, he described certain differences that he felt were typical. Young Mandarin, he wrote, were more clever than young Wood Ducks, were the first to find the best feeding places, and also were more aggressive and pugnacious. In one instance, Lord Grey observed a mixed brood of five Wood Ducks and five Mandarin ducklings, all hatched at the same time under a Wood Duck hen on a cold day in early May. He describes how the rain fell, the wind blew, and a cold night followed. The next day, all five Mandarin chicks were alive and well but only one young Wood Duck survived. The survival of the Mandarin, he concluded, must have been due partly to their constitution and partly to their innate cleverness, which on that cold night led them to gain for themselves the best and warmest shelter under their foster mother, the Wood Duck. He also noted his belief that young Mandarin grow faster and stronger than young Wood Ducks.

Lord Grey was observing two species that are not native to Britain. Now, some sixty years later, we realize that Mandarin in the Far East normally hatch into a colder spring than the Wood Duck does in its temperate North American climate. Information collected at the Wildfowl and Wetlands Trust at Slimbridge in England has shown that Wood Duck eggs and newly hatched ducklings are on average more than sixteen percent lighter in weight than Mandarin eggs and newly hatched ducklings. In Britain, particularly as far to the north as Lord Grey's estate, when both species hatched together on a cold spring day, the heavier Mandarin ducklings would naturally be more aggressive and more alert than the lighter young Wood Ducks, which in their native habitat would normally hatch into a warmer and more hospitable environment. In fact, very young Wood Ducks, given plenty of food in a suitable climate, actually grow faster than Mandarin. Although smaller in size and weighing less at birth, by the time they are ready to fly, young Wood Ducks weigh approximately the same as Mandarin of the same age. Then, as adults, the Wood Duck drake is three to four percent larger than the Mandarin, and the two hens, while slightly lighter than the drakes, weigh very nearly the same. From his observations, Lord Grey concluded that the hardier Mandarin might well establish themselves in a wild state in Britain but that the Wood Duck certainly would not.

In the sixty years since, as the Mandarin became established as a greatly beloved part of Britain's wildlife, the Wood Duck has never really established itself in a natural state. Lord Grey's prophecy was correct, but, as would later be discovered, for the wrong reasons, and from these circumstances developed

Early in this century on his estate, Lord Grey of Fallodon tamed free-flying Mandarin and Wood Ducks as no one else has ever done with either species.

the myth that the Mandarin was more clever, more hardy, and more aggressive than the Wood Duck. Lord Grey simply did not understand that the early spring climate in Britain was more suited to the newly hatched Mandarin than to the more vulnerable young Wood Ducks. Despite his apparent misinterpretation of his observations, there is no denying the affection he brought to the Mandarin and the trust they returned to him.

The first Mandarin to breed successfully and to establish the colony that today numbers in the thousands were believed to have been brought to Britain in 1931 by the world-famous French ornithologist Dr. Jean Delacour. Years ago, Dr. Delacour told me the story of how he had discovered a shipment of Mandarin in the Paris bird market. The birds were in several large bamboo baskets. On examining the appearance of the baskets, he concluded that they had come directly from China, from Fujian or Guangdong, the Mandarin having been trapped or netted there in the wild. The birds were in a sorry state, many having perished from overcrowding. Delacour purchased the entire consignment and had them transported to his house, where he undertook to rescue those that could be saved. He then carried approximately forty-five of the healthier Mandarin to Britain to present to his friend Alfred Ezra, who had a small waterfowl collection on his estate at Foxwarren Park near Cobham in Surrey, southwest of London. Within weeks, all forty-five Mandarin were released, their wings clipped, not pinioned, in order to restrict their flight sufficiently to keep them in the area until the following season when, following their annual molt, they would again be fully feathered and able to fly.

Many did depart the area after their molt, but those that remained bred freely, and it is believed that they and their offspring spread into the environs around Virginia Water on the Berkshire-Surrey border, and to the many ponds of Windsor Great Park and the adjacent forest. Ultimately resident colonies were found in many other nearby areas such as Titness Park near Sunninghill, Ascot Place, and Foleyon Park in Winkfield. For reasons that are not clearly understood, a species that migrates presumably thousands of miles in the Far East has seemingly lost its migratory instinct in Britain. South-central England, with its mild temperatures, ample woods, and wetlands, provides suitable Mandarin habitat that is generally unavailable in other areas of Britain or on the continent. Although sightings of Mandarin on the continent are rare, Dr. G. H. Voorwijk in the Netherlands has Mandarin in his ponds and garden in Heerde in the heavily wooded northeast

TOP The day of the ducklings' departure from the nest, the hen scouts the area to be certain it is safe. Then she returns to the nest to warm the ducklings briefly before calling them out.

ABOVE The last duckling to leave the nest eyes the long drop.

area of the country. He and his neighbors estimate that several hundred Mandarin nest and raise their young in this district. Visitors to his popular gardens see numerous nesthouses hung in the trees near his beautiful villa and watch Mandarin walk from the pond to the side door of his house where they eat from his hands.

There is no written record of the precise date of the Delacour-Ezra release, the exact quantity of birds released, or the number of isolated colonies that may have combined with the Ezra group. Yet there is little disagreement that this group of birds—direct from the forests and waterways of China, strong enough to survive the shipment when others perished—are the hardy stock that produced the present thriving population in Britain, and that the man responsible for their delivery to England was Dr. Jean Delacour.

While I was an undergraduate at Cambridge University about to embark on a career in engineering and already intrigued by waterfowl, I can recall setting off to study the wintering birds in the south of England on Virginia Water Lake. Early one morning while stalking a large flock of ducks, I saw close-by, in a flash of color from under the rhododendrons, a small duck that I was unable to identify. Nor could I find it in any of the books on British birds of the day. Soon after this experience, I met Derek Godwin, another young ornithologist, who suggested that the duck might have been a Mandarin since he had once seen some in that same area many years earlier. A visit

Young Mandarin hide in the grasses on the shoreline of Virginia Water Lake.

A drake dries his wings, clearly displaying his bright plumage: the silver-edged primaries, white-tipped secondaries, pointed scapular, or shoulder, feathers, and prominent sail feathers.

to Sir Peter Scott's Severn Wildfowl Trust at Slimbridge near Gloucester confirmed that the duck I had discovered at Virginia Water was indeed a Mandarin. Of all the ducks at Slimbridge, I was later to write, "My favorite above all was the Mandarin. Both male and female were the most marvelous creatures I had ever seen."

While at the university, I wrote essays about the Mandarin which won the Alfred Newton Essay Prize in 1949 and 1950. The judge for the contest was Professor David Lack of Oxford University, who believed in the research I was engaged in and encouraged me to continue my studies of the Mandarin.

Thus began my lifelong interest in the Mandarin in Britain and in the Far East, which included correspondence with those few individuals who could cast light on the Mandarin in those remote and isolated areas. At that time, many of my colleagues in Japan and China indicated that its population was rapidly decreasing. This led me to believe even more in the importance of taking the Mandarin more seriously in British wildlife circles, which meant having it accepted on the British Ornithologists' Union's official British and Irish List.

In 1952, A. & C. Black of London published my book, *The Mandarin Duck*, the first ever written about the Mandarin. In addition to describing for the first time the natural history of the birds in Britain, the book attempted to relate the history, distribution, and habits of the Mandarin throughout the Far East and also sounded an alarm about the Mandarin's endangered status in its native habitat. The intricate courtship ritual of the Mandarin held great fascination for me. Pencil sketching had long been a hobby of mine, and I tried to apply it to depict all aspects of the Mandarin's activities including courtship. I like to believe that this presentation of the Mandarin to the British public helped to stimulate further investigation of this rare species, which, in 1971, resulted in the Mandarin's being officially placed on the British and Irish List. The species is now protected throughout the United Kingdom.

Twenty years later, another young man with a mission, Andy Davies, came upon the scene. His early interest in wildlife photography caused him to become involved with the Mandarin before it was taken seriously. As Davies recollects, the Mandarin was considered something of a curiosity by serious ornithologists, and his admittedly youthful and amateurish interest in them was somewhat ridiculed. By 1975, after a series of exciting and often frustrating experiences photographing Mandarin at their nest sites, he began what was to become one of the longest-running studies of wild Mandarin ever undertaken.

From 1975 to 1990, while working full-time at another job, he spent almost every available hour studying the natural history of the Mandarin in the area near his home in Surrey. His primary transportation was his bicycle, which he loaded with gear for his trip to the farmland areas he

TOP RIGHT The female's secondary feathers have white tips and several distinctive white fingerprints rarely found on the drake.

CENTER A detail of the drake's scapular feathers shows the dramatic contrast of black and white.

RIGHT A drake in nuptial plumage parades in what is called "full sail."

had under surveillance, frequently an hour away. His work with the Mandarin often began before dawn and ended with his return home after dark, with a full day at his regular job sandwiched in between. The results of his pioneering study of wild Mandarin soon began appearing in British journals and news reports. Davies was fortunate, as few observers have been, to watch the birds from their early spring courting through their nest search, and their laying, incubation, and hatching of eggs. Finally he observed the most rarely seen behavior—newly hatched ducklings jumping from their nests. He was one of the very few students of the Mandarin able, in many instances, to follow them through an entire season.

One of Davies' primary studies concerned nesting Mandarin. In addition to observing them at their natural tree-cavity nests, he was one of the first to study the Mandarin's use of man-made nestboxes in the wild. The latter situation greatly facilitates research since in the boxes the hen, her eggs, and her ducklings are easy to observe. He was the first to identify the Mandarin as an intraspecific "dump nester," a duck that lays its eggs in the nests of other Mandarin, as contrasted with birds that lay their eggs in nests of other species (interspecific dump nesting). Early writers who described Mandarin in the wild observed them nesting only in natural and mostly inaccessible tree cavities, which made close scrutiny almost impossible. The phenomenon Davies described is fairly common, particularly among tree-hole nesting ducks. For those ducks in the wild that require properly sized and located tree holes for nesting but find the supply limited, dump nesting naturally would be more common than for those ducks that need only a flat spot on the ground to build a nest and are not forced to borrow space from another duck. Davies' study of Mandarin over nearly twenty years, a period during which he

Lush vegetation lines a pond where Andy Davies erected nest-boxes and pursued his pioneering study of the Mandarin.

practically lived with them in the wild, contributed other new information on the Mandarin, such as a nesting hen's nocturnal feeding forays, observed while Davies camped beneath the tree where she was setting. He also confirmed in wild Mandarin the behaviors described in older reports of Mandarin studied in semicaptivity.

Less than thirty miles north of Davies, Sir Christopher Lever was observing Mandarin in prime habitat in Windsor Great Park. In 1977 he published his six-hundred-page *Naturalized Animals of the British Isles*, which he dedicated to the Mandarin duck, the "most beautiful of birds." Regarding the advantages and disadvantages of introduced species, Lever concludes in his prologue, "And surely no one could object to the introduction of harmless and beautiful animals which add to the attractions of the British countryside; such a one is the lovely mandarin duck, to which this book, with admiration and affection, is dedicated." In 1990 he wrote for the Shire Natural History series a small book about Mandarin illustrated with color photographs. He is regarded as an authority on the history of the species.

In 1990, Dr. Janet Kear, then an Assistant Director of the Wildfowl and Wetlands Trust in Slimbridge, published *Man and Wildfowl*, in which, from her own observations, she advances convincing explanations for the success of the Mandarin in establishing itself in Britain and the relative failure of the Wood Duck to do so under similar circumstances over almost exactly the same thirty-six-year period. She describes how the Wood Duck in Britain lays its first eggs on or about March 20, whereas the Mandarin lays its first eggs on April 15, over three weeks later. This schedule suits the climate of the Wood Duck's native range, but means that the Wood Duck in England sets on eggs and hatches ducklings when temperatures are still cold. Dr. Kear notes that, as a result, not only do Wood Duck eggs hatch earlier, when temperatures are cold, but, to make matters worse, the young Wood Duck is typically sixteen percent lighter at hatch than the Mandarin and thus loses body heat faster. Her conclusion is that the genes of the Wood Duck have adapted it to the circumstances it finds in North America, which is closer to the equator than Britain, and therefore the Wood Ducks, smaller at birth than Mandarin, are poorly suited to the wet and cold of Britain's early spring. Dr. Kear also describes the greater difficulty of feral wildlife such as the Wood Duck in adapting to the wild after many generations in captivity, compared with the truly wild Mandarin imported directly from its natural habitat in China.

From the writings of these and other dedicated individuals is emerging a far clearer and more complete picture of the natural history of the Mandarin in Britain. The distinctive plumages of the Mandarin drakes and hens in Britain appear identical to those of birds in the Far East or in the United

In 1971, in recognition of the stable population of Mandarin in Great Britain, the species was placed on the British and Irish List.

States. At one time there was some speculation that the white edging on top of the drake's copper sails might be unique to the British birds. In fact, all drakes show the same white trim on top of each sail, with an edge of black at the front and a glossy blue below the quill.

In Britain, as throughout its habitat in the Far East, the Mandarin is a bird of the forest, though the trees in Britain are mostly in manicured parklands often patrolled by gamekeepers or wardens, and in farmlands, orchards, and woodlands with meandering streams or quiet ponds lined with beech, oak, and chestnut trees. The Mandarin in Britain have adapted to a civilized and intimate habitat that seems far different from the wilderness areas of Ussuriland in Russia, China, and even Hokkaido in Japan. Perhaps in ancient times, the shrines and palace gardens of China and Japan conditioned the Mandarin to accept the hospitality of the British countryside.

Wherever Mandarin live, they are tree-hole nesting ducks that lay relatively large clutches of eggs. In Britain, however, Andy Davies found that a normal clutch size is fourteen eggs, compared with reports of ten or fewer in the Far East. In addition to the Mandarin's dump nesting, he also notes the existence of what he identifies as "drop nests" in which eggs are deposited by several hens, nests with a scruffy, soiled, and abandoned appearance. Hens in such nests never cover the eggs with shavings or down, and the eggs are not incubated. This same phenomenon of drop nesting occurs among both Mandarin and Wood Ducks in California.

Food for the British Mandarin is much the same as in the Far East. The mast of forest trees such as oak or beech is favored, as are the small European sweet chestnuts. Seeds of many plants are a mainstay of the Mandarin, and in Britain specific mention is made of buttercup, clover, and pink pondweed. Insects of all kinds, as well as small frogs, provide protein in the spring. Mandarin are frequent visitors after the harvest to stubble fields.

Some observers in Britain believe that without supplemental feeding by humans the Mandarin could not long survive. Others believe that certain areas in the south of England possess sufficient food and nesting potential to support the Mandarin, but that throughout the rest of Britain the shortage of proper habitat and food has kept the Mandarin concentrated in the Surrey-Sussex areas.

Perhaps the most important aspect of Andy Davies' research dealt with the population and distribution of Mandarin throughout Britain. Over a period of three years, he examined and cataloged old records, and investigated unpublished personal communications from individual observers and bird clubs. Lastly, and less conventionally, he interviewed landowners, gardeners, gamekeepers, fishermen, and ornithologists. In the end, he had nearly one thousand original source documents, all of which were painstakingly plotted into areas of approximately six square miles. He summarized his findings in a fifty-one page document based principally on records from 1970 to 1986, years not covered by previously published maps. Although he recorded and plotted more than one thousand

BELOW Bert Winchester, using an eight-foot duck trap of his own design, assisted Andy Davies in the most successful Mandarin trapping and banding program ever undertaken in any country where the birds are resident. **BOTTOM** More Mandarin have been banded and more nest-boxes have been provided for the ducks in England than in all of the regions of Asia.

sites at which Mandarin had been seen, the map shows that almost seventy-five percent of the Mandarin in Britain are in the two areas where he conducted his fourteen-year nest study program.

To interpret this vast accumulation of data, Davies then had to turn to the results of a Mandarin trapping program he had begun in October 1985 with an older and more experienced birder and trapper, Bert Winchester. Prior to 1985, the total number of Mandarin caught and banded in Britain was only sixty-one birds, mostly female, which had been captured by hand when they were nesting. By the time Davies and Winchester had trapped and banded five hundred Mandarin in the water and on the shoreline, only two banded birds in ten were being retrapped, and only one in ten birds observed throughout the area was banded. This was interpreted to mean that in the area where trapping occurred, the total population was between twenty-five hundred (two in ten being retrapped) and five thousand (one in ten being observed). Further, their trapping indicated that many more Mandarin were in any given area than were being counted. Therefore, it appeared that the conventional way of counting by sighting wildfowl from September to March drastically underrecords Mandarin and is not appropriate for estimating this secretive species. Based on this radically different approach, and having plotted the percentage of Mandarin in the trapping area alone, Davies extrapolated a British population of seven thousand to thirteen thousand individuals.

The vibrant coloring of the British Mandarin reflects the vitality of those native birds imported from eastern China.

This was a startling calculation. Not only was Davies' approach designed to accommodate the secretive and retiring nature of the Mandarin, but it also was based upon hard data gained by trapping, banding, and retrapping banded birds. This pragmatic method produced a total population that was more than twice that previously projected by the experts.

Those of us familiar with the character of both Mandarin and Wood Ducks tend to accept the population figures projected by Davies and might well feel that even these calculations undercount the Mandarin in Britain. Andy Davies has convinced most Mandarin watchers that there was in 1986, and undoubtedly still is, a substantial and slowly increasing population of Mandarin in Great Britain. The status of the species in Britain ultimately could have great significance for the Mandarin in Asia should the depletion of the birds and their native habitat continue.

That Mandarin are in Britain at all—and that they have been studied and cared for to create today's thriving population—is a tribute to all those who have accepted the species as one of their own. [C.S.]

The Mandarin have no better supporters in the world than the British, who see the species as a talisman for the conservation of all living creatures.

URAWAKAKI
UME O KUGURITE
SHIJUKARA
HARU O SAKASA NI
KOSU KOKOCHI SEN.

OKU SHIMO NO
UE NI MO YUKI O
KASANEKITE
KIREBA KIEMU TO
OSHIDORI NO NAKU.

Little chickadee,
twisting about the plum tree,
glowing with youth,
must want to be upside down
when New Year's Day arrives.

A layer of frost,
and on top of that a snow,
slowly deepening.
"We will perish if we part,"
the mandarin ducks are crying.

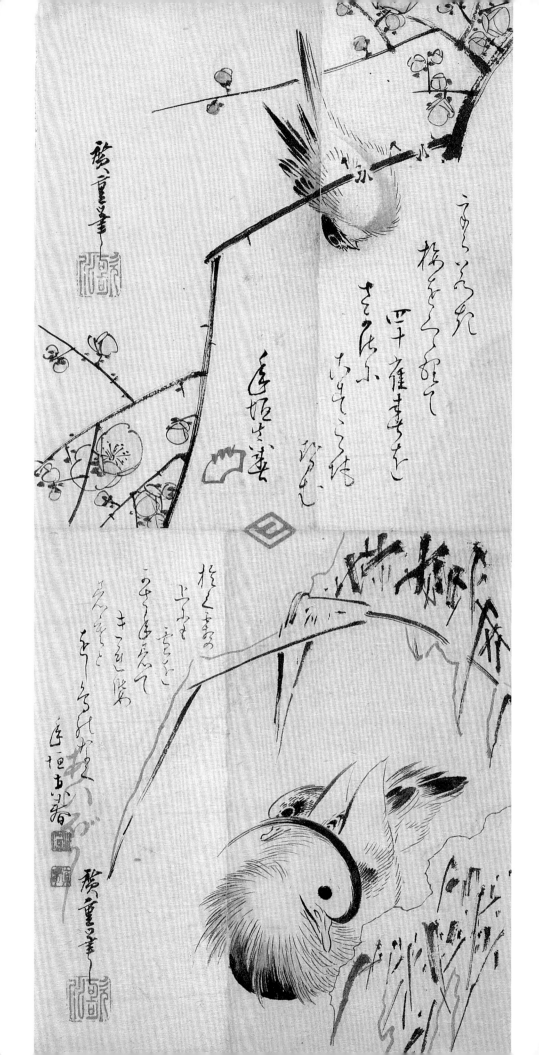

CHAPTER FIVE

THE MANDARIN IN ART AND LITERATURE

*'Tis winter and
snow time
The cold wind
blows hard
And roughens
the water;
But, peacefully,
the Mandarin
swim on.*

—CHINESE ODE
(CIRCA 400 B.C.)

To BEHOLD THE MANDARIN TODAY is to be in the presence of a species that serves as a living link to the ancient cultures of China and Japan and to the founders of the religious faiths of nearly half of the world's inhabitants. From the earliest centuries of the Mandarin's known history in the Far East, it has been admired for its beauty and praised for its virtuous qualities. Poets, storytellers, and fablers have constructed verses, odes, and allegories around the Mandarin and its habits. Visual artists have woven and embroidered the Mandarin's image into textiles, painted it on canvas and paper, and sculpted it in innumerable materials both precious and humble.

Like the crane, which symbolizes happiness and longevity, and the array of other animals highly regarded throughout the history of Japan and China, the Mandarin represents particular enduring qualities: loyalty, marital fidelity, and conjugal felicity. This centuries-old reverence for the species originated in the belief that the Mandarin drake and hen are monogamous and that they display an affection for one another that is unusual among ducks. Adult birds form pairs early in the breeding season and sometimes by the time the birds reach their wintering places. To many early observers of Mandarin in temple ponds or other prominent sites, these pairs seemed

OPPOSITE A poem by Toshigaki Maharu extolling the loyalty of the Mandarin drake and hen is an elegant and integral part of Japanese artist Hiroshige's *Titmouse on a Flowering Plum and Mandarin Ducks in Snow*, a drawing in ink and color on paper from the 1830s. (Courtesy of the Museum of Art, Rhode Island School of Design, gift of Mrs. John D. Rockefeller, Jr.)

175

to have been the same pairs formed the season before. According to an ancient Chinese belief, a Mandarin that loses its mate pines away and dies. These interpretations of Mandarin behavior have been the source of many exquisite depictions of the species from the hand of the writer and the hand of the artist, historical examples of which are presented on these pages.

Although mated pairs of all species of perching ducks preen each other's head and neck feathers, this practice between the striking Mandarin drake and the exquisite hen undoubtedly reinforced the impression that they were a devoted couple caressing each other. Observers centuries ago also must have been captivated by the frequency of a Mandarin pair's courtship displays and by their tendency to drive off intruders, as if they were asserting their affection and commitment. Even today, one of the most charming characteristics of a paired Mandarin drake and hen is the way they stand closely side by side, perched on a branch over the water or under an overhanging riverbank during rain or snow.

Artists in the West, particularly British artists who have benefited from the opportunity to observe the Mandarin resident in England, have also rendered the species in works of art in a wide range of media. Influenced by the symbolic significance of the Mandarin in Asia, many Western artists maintained the tradition of presenting the loyal Mandarin drake and hen together. Artists in China and Japan had been living amid the Mandarin and using its image in their work for many centuries before Europeans first viewed the species in its native lands. Kaempfus, the Dutchman who was among the early Europeans to enter Japan, described the Mandarin in his 1723 *History of Japan*, referring to the Mandarin as the Kinmodsui. Impressed with the appearance of the male, he wrote: "Its feathers are wonderfully diversify'd with the finest colours imaginable; about the neck and breast chiefly they are red. The head is crowned with a most magnificent topping. The tail rising obliquely, and the wings standing up over the back in a very singular manner, afford to the Eye a sight as curious as it is uncommon." The accompanying simplistic, chiefly documentary illustration, lacking the extraordinary subtlety

A pair of Mandarin are devotedly intertwined in a small delicate sculpture carved in jade during the Ch'ing dynasty (1644–1911), a period when the arts flourished in China. 6¼ inches long. (Courtesy of the Asian Art Museum of San Francisco, The Avery Brundage Collection, B60 J442)

and beauty found in Japanese and Chinese art of the time, highlights the Mandarin drake, not surprisingly with the hen close-by.

The duck that today is known as the Yuen Yang in China and the Oshidori in Japan was given the name Mandarin by the English in the 1700s. The word *mandarin* comes from the Sanskrit word *mantrin*, meaning "counselor." The word was first recorded in the late 1500s as being applied to Chinese officials. Other references suggest that the original Sanskrit term was used by foreigners to describe the handsomely attired senior officials of the Chinese government and also came to refer to the yellow color of their silk robes. So perhaps the use of the name *Mandarin Duck* was a way of saying "the splendid yellow duck."

Some of these early travelers to the Asian continent, clearly fascinated by the exotic appearance of the Mandarin, introduced imported stock to England. These captive birds were the source for the depictions of the Mandarin by early ornithological illustrators. A plate by George Edwards in his *Natural History* of 1747 was drawn from life in the gardens of Sir Matthew Decker in Richmond, England. Although this early work is inferior by modern standards, it is interesting because it is the first indication of the introduction of the Mandarin to England. Like Edwards, other ornithological artists, such as Georges-Louis LeClerc de Buffon, illustrated the drake alone, but with more graceful results. John Gould, for his models,

A pair of Mandarin idle along the shoreline in a six-part folding screen, in ink, color, and gold, by sixteenth-century Japanese artist Genga. 64 5/8 inches high by 129 inches wide. (Courtesy of the Freer Gallery of Art, Smithsonian Institution, 71.2)

John Gould studied the Mandarin in England during the period when Audubon observed the Wood Duck in North America. His lithograph appeared in his seven-volume *Birds of Asia*, published in the mid-1800s. (Courtesy of the California Academy of Sciences, Special Collections)

visited the Mandarin at the London Zoo. From these sketches, he created his much-admired rendering of a Mandarin pair rather than the Mandarin drake alone.

Even before the Mandarin had attained its symbolic status in Asia, it was among the animals and plants that appeared in verses intended to impart wisdom. The earliest known references to the Mandarin date to the fifth century B.C., the time of Confucius, who urged his followers to collect such odes and songs because they would promote a harmonious society and encourage wisdom and peace among its citizens. "A knowledge of the songs," Confucius counseled, "enables us to incite other men to desirable courses, helps to observe accurately their innermost feelings, and to express our own discontents, to do our duty to parent and prince, and finally to widen our acquaintance with the names of birds, plants, and trees." These odes and songs, widely known and quoted by educated people during Confucius's time and long after, were recited on special occasions. One of these may have been used as a speech of blessing to a guest.

> Mandarin ducks were in flight,
> We netted them, we snared them.
> Long life to our Lord
> Well may blessings and rewards be his!
>
> There are Mandarin ducks on the dam,
> Folding their left wings.
> Long life to our Lord,
> Well may blessings for ever be his!
>
> When there is a team of horses in the stable
> We give it fodder, give it grass.
> Long life to our Lord,
> May all blessings nurture him!
>
> When there is a team of horses in the stable
> We give it grass, give it fodder
> Long life to our Lord
> May all blessings safely bind him!

The ducks' folding of their left wings portends blessing heaped upon blessing, a connection that occurs in other verses of the time. The expression could have been given currency by the Mandarin's habit of pointing to its folded wing during display while lifting one wing after the other to arrange its plumage.

To Chinese Buddhists as well, the image of the Mandarin was used to impart wisdom. Beginning in about 200 B.C., the Mandarin was viewed as a model to mankind of the wisdom of kindness and compassion. Many are the stories and legends connecting the Mandarin with a "Great Enlightened One." When Buddhism came to Japan in the sixth century A.D., so did the Mandarin's symbology. One of the Wisdom Buddhas, Amida Buddha, was said to have assumed the form of the Mandarin Duck to teach lessons of compassion.

Many examples of secular literature survive as a tribute to the Mandarin's symbolic significance in China, such as "The Beautiful Woman" by Tu Fu, dating from the Tang dynasty (A.D. 618–907).

TOP In his *Natural History of Birds*, dated 1747, Englishman George Edwards included an engraving of the Mandarin drake. His written description, in both French and English, refers to the species as Cercelle de la Chine, or Chinese Teal. (Courtesy of the Smithsonian Institution Libraries, OPPS no. 95-2526)

ABOVE Georges-Louis LeClerc de Buffon published numerous editions of his books on natural history. His hand-colored plate of the Mandarin drake is in his *Histoire Naturelle des Oiseaux (Natural History of Birds)*, 1770–86. (Courtesy of the California Academy of Sciences, Special Collections)

My husband holds me in light esteem,
But his new mistress seems as beautiful as jade.
Even the morning glory has its passing hour.
The Mandarin duck and drake do not roost apart,
But wrapt in his new favorite's smile,
How can he hear his old love's sighs?

In a similar vein, "A Song of Chaste Women" by Meng Chiao holds the fidelity of the Mandarin drake and hen as a model for human values and behavior.

The Wu-tung trees grow old together,
The Mandarin duck and drake pair for life,
Even so the chaste woman prides herself on following
her husband to the tomb,
And throws away her life.
I vow no waves shall ruffle the surface of my passion
Which is still as the waters of an ancient well.

The earliest existing Chinese painting of the Mandarin dates from the Sung dynasty (A.D. 960–1260). Tsao Chung depicted in black ink part of a stone and a sunflower, beneath which rest a pair of Mandarin. He was a native of Chengdu, which is now outside the range of the Mandarin. Yet there seems little doubt, from the quality of his work, that the artist was familiar with the Mandarin. "Other painters can produce an accurate resemblance of the flowers they paint," goes an account of the painter's life, "but Tsao Chung not only produces an accurate resemblance, but hands over to you the very soul of the flower along with it." And so he has done with the Mandarin.

During the Yuan dynasty in China, from the late thirteenth to the mid-fourteenth centuries, nobles wore robes bearing patterns of birds and animals, one of the favored images being that of the Mandarin. These woven or embroidered designs were usually confined to an ornamental collar or to a square on the chest or back of the garment, and a band across the knees. The court of the Ming dynasty continued to use bird and animal patterns on robes but at the end of the fourteenth century began to regulate the use of these designs to reflect status and position. Birds were assigned to civil officials and mammals to military officers. Birds, it was believed, symbolized literary elegance, while mammals bestowed fierce courage on their wearers. The laws specified that officials of the sixth and seventh ranks were to wear the egret or the Mandarin on their robes. As the laws were changed, a single species was specified for each rank, the Mandarin being reserved for the seventh grade. The Mandarin retained this rank throughout the centuries until the Chinese Revolution ended dynastic rule in 1911.

Thus, for over five centuries, one of the main and most interesting decorative uses of the Mandarin in China was on these squares worn by officials and their wives. Although ranking was specified, style was not, and a wide variety of treatments may be found on these textiles, always executed with exquisite and intricate craftsmanship. The fibers used to embroider or weave the squares were usually silk and occasionally were wound with gold or silver strips. Squares from the seventeenth century bearing renderings of the Mandarin and other images had traces of peacock herl.

Of great significance to the appearance of the Mandarin in Japanese art was the reign of Emperor Shomun in the late eighth century A.D. He is most famous today for his encouragement of Buddhism and the arts of Japan and as the builder of the Todaiji Temple in Nara, still the largest freestanding wood building in the world. Emperor Shomun also was a great art collector and admirer of China. He not only sent teams of envoys to collect some of the finest artifacts of the day but also brought Chinese craftsmen to Japan to introduce their skills. At Emperor Shomun's Shosoin, or Treasure House, the Mandarin abounds in religious artifacts and as decoration on secular objects used in the emperor's court. Few portrayals of the Mandarin exist before this time, which makes the works preserved in Shosoin of immense value to the significance of the Mandarin in Japanese art even to this day. Upon the emperor's death, his collection of thousands of art objects was sealed in the treasury. It was not until shortly before World War II that objects in the collection were unpacked and cataloged. So well built was Shosoin that it survived earthquakes unscathed, and so well ventilated was the collection that the objects are preserved in pristine condition. Although work on the collection continues, a new selection of pieces is displayed for the public for a few days each year.

Lotus blossoms surround a Mandarin drake and hen floating above stylized mountains in a Chinese silk tapestry, or *k'o-ssu*, created during the Ming dynasty (1368–1644). 23 1/8 inches high by 15 3/8 inches wide. (Courtesy of the Board of the Trustees of the Victoria and Albert Museum)

The art objects preserved in Shosoin demonstrate the level of technology and craftsmanship introduced to Japan from China and the influence of designs used in China during the Tang dynasty, regarded as China's second

Golden Age. Among the objects in Shosoin are huge bronze mirrors that incorporate images of the Mandarin. Connected with the ceremonies at the emperor's Todaiji Temple, they were hung from the roof by ropes tied to heavy bosses on their reverse sides. When the temple was lit by thousands of candles, the mirrors reflected the light and created the illusion of a starlit heaven above. The backs of the mirrors, which would have been seen only by priests and monks, contain marvelous engraved scenes of the Buddha surrounded by seas and mountains full of recognizable birds and animals, including Chinese dragons and phoenixes. Among these creatures are both stylized and realistic Mandarin. Considered sacred, the mirrors were protected in silk-lined boxes made of cedar.

Chinese mirrors from the Tang dynasty continued to influence Japanese art over the succeeding centuries. Small mirrors, sometimes as intimate in size as three inches in diameter, were intended for men and women to use in the home. Cast in bronze, they bear on their reverse sides decorative designs of stylized animals and birds such as Mandarin and cranes. Mirrors like the eight-lobed one from the twelfth century illustrated here also had a spiritual significance. The owner of the mirror polished the reflective surface to ensure that his or her soul remained clear and bright.

In Japanese screen paintings, Mandarin swim on ponds or stand on the shore, the drake alongside the hen, continuing the symbolism of showing the faithful male and female Mandarin together. Folding screens, whether consisting of only two attached panels or composed of as many as twelve panels, were functional objects as well as works of art. These *byobu*—meaning "protection from wind"—were, in a sense, movable walls used to partition space. The scale and format of the screens made them particularly well suited to the depiction of the Japanese landscape and the plants and animals within it. Using ink and colored pigment, often including gold, on paper, painters such as Genga treated the multipart screens as a continuous surface. In the screen by this sixteenth-century artist, a pair of Mandarin are at the center of a scene that contains a number of identifiable birds, plants, and flowers.

Rather than forming a continuous scene, other screen paintings are composed of twelve panels devoted to the flora and fauna associated with the twelve months of the year. In the Edo period, from

A pair of Mandarin face each other on the back of an eight-lobed bronze mirror from the Fujiwara period (early 1100s) in Japan. Measuring 5½ inches in diameter, it was held in the hand rather than hung on the wall. (Courtesy of the Asian Art Museum of San Francisco, The Avery Brundage Collection, B65 B56)

the seventeenth through the mid-nineteenth centuries, these screens elegantly combined the arts of painting, poetry, and calligraphy into a form called *waka-e*, meaning "poetry painting." Yamamoto Soken's *Flowers and Birds of the Twelve Months* commences the year with a pair of Mandarin poised along a stream, the bare branches and dusting of snow revealing the time of year. Directly inscribed in calligraphy on the panel for each month are two poems by Fujiwara Teika, one on the bird shown, the other on the plant.

Imagery, poetry, and calligraphy also merge in the *kacho-ga*, or bird-and-flower work, of Hiroshige, one of the most accomplished artists of the Edo period. Although best known for his depictions of the city of Edo (Tokyo) and the pursuits of its people, Hiroshige created elegant designs for several hundred woodblock prints of plants and animals. Among the drawings and prints of falcons, sparrows, cranes, and kingfishers are a number of works with Mandarin ducks paired according to tradition and shown during different seasons of the year. Calligraphy carefully placed within each balanced composition offers haiku, the seventeen-syllable verse, or other types of poetry, such as thirty-one syllable *kyoka*. These works by Hiroshige, like those of Chinese painter Tsao Chung, rather than serving as scientific description of nature, are refreshing interpretations of the very essence or spirit of the Mandarin.

A pair of Mandarin along a tumbling stream begin a series of paintings, in ink and colors on silk, depicting the months of the year. The set of two six-fold screens by Japanese artist Yamamoto Soken, one of which is shown here, was commissioned during the Edo period (1615–1868), probably by a courtier for use as a gift. 44 1/2 inches high by 17 3/8 inches wide, each panel. (Courtesy of the Asian Art Museum of San Francisco, The Avery Brundage Collection, B60 D82+B)

The Mandarin in Haiku

The drake and his wife
paddling among green
tufts of grass
are playing house.
— ISSA

Snow falls lightly
on the wings of the mandarin ducks:
The stillness!
— SHIKI

Mandarin ducks;
a weasel is peeping
at the old pond.
— BUSON

Evening snow falling
a pair of mandarin ducks
on an ancient lake.
— SHIKI

The morning tempest
sees even mandarin ducks
go separate ways.
— RIHO

Take an angry man
and show him mandarin ducks
— they never break up.
— ANONYMOUS

This Japanese garment of silk and satin, a *furosode* dated 1750–1850, was worn by a bride as part of her wedding costume. Embroidered among the stylized blossoms on the back are two pairs of Mandarin, shown here in detail. (Courtesy of the Museum of Art, Rhode Island School of Design, gift of Marshall H. Gould)

Haiku are among the most characteristic and charming examples of Japanese poetry. Delicate and magical in effect, they encapsulate more than one idea in a succinct combination of words. Many haiku poets chose the Mandarin as the central image (see opposite). The most famous haiku on the Mandarin is by Yosa Buson (1716–83):

> Mandarin ducks,
> Epitome of beauty
> The winter wood.

It was composed at Ruo-anji Temple in Kyoto, noted for its rock garden and small pond by the entrance, called *oshidori-ike*, or Mandarin Duck pond, which used to be frequented by Mandarin.

The Mandarin also has a most honored place in the folklore of Japan, and various legends have, as a recurring theme, a drake or hen returning to its dead mate or appearing in a dream. As is typical of Japanese mythological narratives, those about animals are presented in the form of a transformation. One of the most touching stories is the legend of Sonjo, who passed by a marsh on his way home after failing to procure game for supper. He saw a pair of Mandarin on a nearby river. Although he knew that the Mandarin were emblems of conjugal affection, he shot the drake and took the bird home. That night in a dream, a beautiful woman stood beside his bed weeping bitterly and beckoned him to return to the marsh. The next day, as he stood on the riverbank, he saw the female Mandarin swimming alone. Rather than flee from him, she swam toward him and stepped out of the water at his feet. Suddenly, as if in grief, she tore open her own breast and died before the hunter's eyes. The hunter was so stricken by the sight that he renounced hunting and became a priest. When he died, the people of his village, touched by his sad story, set up a shrine, which can be seen today, at Utsonomiya, north of Tokyo. A stone is inscribed with the story, and a cedar stake records the Mandarin's name.

In Japan in particular, where Mandarin can still be seen in village streams and on temple ponds, the ancient symbolism of the Mandarin persists in many forms. The legends of the Mandarin are kept alive in the annual Oshidori festival organized every year by a society in Utsonomiya called the Oshidori Zuka Ai Gokai. And even today, a young couple may receive as a wedding gift a pair of small, gilded pottery bells in the shape of Mandarin— a gesture that blesses their fidelity and devotion. It is this enduring regard for the Mandarin that may help ensure the preservation of the species throughout its Asian homeland.   [C.S.]

Mandarin Ducks on an Icy Pond with Brown Leaves Falling, a woodblock print from the 1830s by Japanese artist Hiroshige, is graced by a haiku: "For mandarin ducks / thin ice is a wedding cup / for a thousand years." (Courtesy of the Museum of Art, Rhode Island School of Design, gift of Mrs. John D. Rockefeller, Jr.)

CONSERVATION

THE CONCEPT THAT WE ARE DEPENDENT on the natural world for our own survival," writes Sir Peter Scott in the foreword to this book, "is the beginning of all wisdom." After thousands of years first consuming and, in more recent times, pillaging the natural world, mankind has begun, at last, to conserve it. The significance of the problem was underscored when astronauts traveled far enough into space to see the planet as a finite sphere with frighteningly limited resources rather than the limitless expanse of land, water, and sky that it appears to be from its surface. Advances in science, particularly in the life sciences, confirmed that life on this planet cannot long be sustained unless the pollution and other misuses of the land, air, and water are halted.

The Wood Duck of North America and the Mandarin of Asia, more than most species, depend on very special resources for their survival. For if the ancient trees and sheltered wetland habitat essential to nesting and brooding were to disappear, the extinction of both species would be inevitable. With forests and wetlands threatened worldwide, it is urgent to embark on a worldwide effort to save them and all of the species they support. "As the twentieth century draws to its close," Sir Peter Scott writes, "the twenty-first has to become the Century of Global Conservation."

Land, then, is not merely soil; it is a fountain of energy flowing through a circuit of soils, plants, and animals.

—ALDO LEOPOLD,
A SAND COUNTY ALMANAC
(1949)

OPPOSITE Efforts to sustain and even to increase the Wood Duck population will depend on preserving habitat like this ancient bald cypress swamp while maintaining effective hunting regulations.

THE NORTH AMERICAN WOOD DUCK AND ITS REMARKABLE RECOVERY

TOP AND ABOVE When duck hunting was unregulated, Wood Ducks became so scarce that it was rare to see a hen and her brood or a solitary drake standing guard near the nest site.

The rapid decline in the Wood Duck population in the late 1800s was widely reported by hunters and naturalists throughout the continent. By this time the Passenger Pigeon, which once numbered nearly five billion birds, was already well on its way to extinction. Only a short time earlier, flocks of pigeons would darken the sky for an entire day as they passed overhead, and their nighttime roosts sometimes covered twenty-five to thirty square miles of woodlands. Gone forever by 1914 was a species that might be alive today had anyone known enough and cared enough to come to the defense of what may have been the most abundant bird to inhabit the earth.

Surely the sad spectacle of the demise of the Passenger Pigeon was an urgent call to action to preserve other threatened species such as the Wood Duck. The immediate danger was the wanton and unregulated year-round slaughter of these handsome and delectable birds for commercial and private ends. Market hunting with guns that brought down scores of ducks in a single blast made huge inroads into the population of waterfowl—as did hunting during the spring nesting season. The loss of every nesting hen represented the loss of perhaps ten to twelve ducklings that she was about to produce. The efforts of people acting singly or as a group and the responsiveness of the various levels of government to their citizens were crucial to the recovery of the Wood Duck. The first measures to control the hunting of Wood Ducks took place at the local level. In 1904, for instance, the state of Louisiana, responding to citizen pressure, forbade the hunting of Wood Ducks entirely for five years. Other states followed, until more than twenty of the forty-eight contiguous states were regulating duck hunting before the federal government was compelled to address the problem.

In 1913 the Weeks-McLean Bill became the federal law that placed all migratory birds in the federal domain and specifically prohibited the hunting of Wood Ducks. This law was extended in 1916. The passage of the Migratory Bird Treaty Act two years later, along with a similar act already passed in Canada, effectively prohibited the hunting of Wood Ducks in both countries. In 1941, when the comeback of the Wood Duck seemed assured, hunting in the United States on a limited basis was granted on a state-by-state basis.

Following these actions at the federal level, two other highly successful conservation programs illustrate the cooperation that can occur when the public's interest is sufficiently aroused. In 1934 Congress passed the Migratory Bird Hunting Stamp Act. One of its main architects was Ding Darling, a talented cartoonist, an ardent conservationist, and, as luck would have it, a man skilled in the political arena. In recognition of his efforts on behalf of the act, he was afforded in 1934 the honor of designing the first federal duck stamp. Each year every duck hunter in the United States—a total of over one hundred million from 1934 to date—is required to purchase a federal duck stamp with the purchase of a hunting license. The sale of these stamps has produced almost

BELOW The first federal duck stamp sold to hunters to finance conservation was issued in 1934.
BOTTOM Ding Darling, the father of the federal duck stamp program, created this drawing for the 1934 stamp.

$500 million for the acquisition of nearly forty million acres of wetland habitat and for easements to protect an additional one and one-half million acres for the benefit of wildfowl.

So successful was the federal stamp program that in 1971 California issued the first state duck stamp, and today thirty-eight states similarly require hunters to purchase these stamps. Throughout the nation wildlife artists compete to have their paintings accepted to illustrate these stamps. States use the proceeds in various ways in the cause of conservation-related projects. The fascinating story of duck stamps, their design, and their collection is a tribute to the artistic as well as the practical success of these programs. Perhaps the greatest national honor ever bestowed on the Wood Duck was its selection in 1968 for a United States postage stamp—the first of an annual series calling attention to the country's need to protect its wildlife.

Then in 1937, the Congress passed legislation that many consider the most important ever enacted to conserve wildlife in the United States. The Wildlife Restoration Act, better known as the Pittman-Robertson Act, laid claim to the funds generated by an eleven-percent excise tax on the sale of sporting firearms and ammunition, money that was then distributed to the states for the conservation of all wildlife. Every year, more than thirty million hunters contribute to this fund through their purchases of arms and ammunition. As a direct result, states now own or manage over sixty million acres set aside for the protection of wildlife and for research concerning land and wildlife management.

One little-understood fact is that since game reserves are funded largely, if sometimes indirectly, by hunters, they are thought to exist exclusively for the benefit of hunters. Yet the nonhunting public visits the reserves to enjoy these protected areas and the abundant wildlife in numbers often twenty times greater than hunters. In large part, this is because hunting seasons are limited to only a month or two, and during the other months, the reserves are open to the public.

In support of and in addition to efforts by federal, state, and local governments to protect the environment, private organizations by the score have been formed, many with general conservation goals,

BELOW Following the success of the federal duck stamp program, states initiated their own programs, led by California in 1971.

BOTTOM Federal postage stamps help educate the public about species of wildlife. This 1968 stamp was the first to honor the Wood Duck.

others with more specialized goals. Among the more familiar names are the World Wildlife Fund, with its familiar panda logo designed by Sir Peter Scott, one of its founders; the Wildfowl and Wetlands Trust in Great Britain, which Scott also founded; the Nature Conservancy; the National Audubon Society; and Ducks Unlimited. Although the invaluable accomplishments of these and many other associations are too numerous to describe, selected examples suggest the scope and magnitude of their efforts.

Ducks Unlimited describes itself as "the world's leading private sector waterfowl and wetlands conservation organization." Founded in 1937, Ducks Unlimited has raised, by its own account, over a half-billion dollars to conserve five-million-plus acres of habitat throughout the United States and Canada. Over five thousand separate projects have been generated, and countless wildlife species benefit from the conservation efforts. Habitat preservation is a major mission of the Nature Conservancy in its efforts to save endangered plants and animals. The conservancy has over a half-million members whose financial support supplemented by other donations has enabled the protection of more than five and one-half million acres of forests, marshes, prairies, mountains, deserts, and islands.

The conservation efforts of the government and of most major associations are directed essentially at the large environmental issues of habitat preservation, from which all species benefit. Some organizations, such as the Sierra Club, concern themselves as well, when the need arises, with individual species and their specific habitat requirements. Increasingly, such groups are cooperating with each other. In 1974, the National Audubon Society and the Nature Conservancy, working with the Beidler family in South Carolina, established the Francis Beidler Forest, a magnificent collection of ponds, swamps, and wetland forests, part of the famous 4-Hole Swamp. The forest is owned by the two organizations and is managed by the Audubon Society as a part of a nation-wide system of more than seventy-six wildlife sanctuaries. Within the forest are dense groves of giant cypress that provide outstanding natural-cavity nesting facilities for Wood Ducks. At one time scheduled for logging, the forest was saved by the creative cooperation of the landowner and these two large, privately funded organizations.

TOP Popularizing a species, as wildfowl groups like Ducks Unlimited do in their publications, can be an important step in its conservation. This 1972 watercolor by Basil Ede is considered one of the finest paintings of the Wood Duck.

ABOVE David Maass created the image selected for the federal duck stamp issued in 1974.

ABOVE Although unregulated hunting was the first major threat to the Wood Duck, destruction of the habitat that the species needs for nesting—by draining wetlands and cutting forests—was a less obvious but more insidious threat.

OPPOSITE As part of ongoing efforts to conserve irreplaceable wildlife habitat, the Nature Conservancy and the National Audubon Society acquired the Francis Beidler Forest in South Carolina.

Another cooperative effort, by the U.S. Fish and Wildlife Service, the Illinois Department of Conservation, the Nature Conservancy, and Ducks Unlimited, concerns the Cache River Wetlands Center, an environmental partnership to protect sixty thousand acres. This extensive complex of cypress and tupelo swamps and bottomland hardwood forests harbors a vast array of migratory waterfowl, wading birds, neotropical migrant songbirds, and rare mammals. The massive, ancient cypress trees, fifteen hundred years old, are the oldest living things in the United States east of the Rocky Mountains. Included in this prime Wood Duck habitat is the Frank Bellrose Waterfowl Reserve, named to honor the man known as Mr. Wood Duck throughout the world of duck lovers. The reserve's twenty-one hundred acres are home to thousands of ducks and geese. River Otters fish in the placid waters, and Bobcats hunt among the oak and hickory trees of the uplands. Over three hundred fifty acres consist of swamps and bottomlands, important nesting habitat for Bellrose's beloved Wood Ducks.

Unquestionably, federal legislation limiting the hunting season and the number of Wood Ducks that could be harvested was the major factor in the comeback of the Wood Duck. But individuals with less ambitious yet equally

THE
FRANCIS BEIDLER FOREST
Established by
National Audubon Society
and
The Nature Conservancy
With the cooperation of the Beidler Family
and many generous contributors

DEDICATED 1974

Ecology and Management of the
WOOD DUCK

important goals also have contributed to the preservation of the species. It is to these individuals the Wood Duck owes, in large measure, its successful return from what many believed was the brink of extinction.

One of the earliest of these micromanagers of the species was Alain C. White. In 1918, when the Wood Duck population was seriously depleted, he believed it prudent to import Wood Duck stock from breeders in Belgium. From these imports, he bred and released into the wild more than two thousand young Wood Ducks. David Grice and John P. Rogers, beginning in the early 1950s, in New England, studied and wrote about the Wood Duck in detail and provided new and important information about the ecological needs of the species. Frank Bellrose and Arthur Hawkins have devoted a large part of their lives working in the field, studying and writing extensively about the Wood Duck. They also developed safe and functional nestboxes that attracted Wood Ducks. In 1994 Bellrose coauthored with Daniel J. Holm the magnificent *Ecology and Management of the Wood Duck*, a book of nearly six hundred pages that draws on more than fifty years of experience.

Frederic Leopold, while running a successful business, found the time to erect dozens of nestboxes on his estate above the Mississippi River. He studied every aspect of Wood Duck behavior, beginning with nest selection and continuing until the hens and their chicks tumbled down the steep cliff at the edge of his property to cross the Mississippi and look for brooding grounds on the other side. Leopold's detailed observations of hundreds of Wood Duck hens and many thousands of ducklings are described in his *Thirty-five Year Study of Wood Ducks on the Mississippi*.

A small group of retired farmers in central California has erected and maintained several hundred Wood Duck nestboxes along the Stanislaus River and has kept records of the ducklings produced from these nestboxes. Stephen Simmons, while working full time as a high school teacher in nearby Merced, also in central California, has erected and maintained, often with the help of his students, nearly two hundred nestboxes on nearby ranches, some nearly fifty miles from his home. For several years he undertook one of the most time-consuming research jobs of all—keeping records of the exact day, even the exact hour, that every nest he supervised produced newly hatched ducklings. Since young ducklings remain in the nest for only

TOP Frank Bellrose, coauthor of this invaluable book, was named Wood Duck man-of-the-year in 1988 by the North American Wood Duck Symposium for his years of work on behalf of the species. ABOVE Frederic Leopold was similarly honored at an earlier symposium.

LEFT Frank Bellrose visits Indian Meadow Ranch and, with Lawton Shurtleff, examines a natural cavity where Wood Ducks nest near the cabin.

BELOW Volunteers in Minnesota publish an informative newsletter to share research on the Wood Duck and to create enthusiasm for conservation programs involving people of all ages.

BOTTOM The National Audubon Society selected the image of young Wood Ducks at Indian Meadow Ranch by award-winning photographer Carl R. Sams II for the cover of one of its calendars.

approximately twenty-four hours after hatching, he had to be present at each nest at the precise moment of hatching to set up and climb a ladder to inspect the nest. Too early and it was a wasted trip; too late and the ducklings were gone. His goal was to staple tiny web tags to the feet of each duckling. Simmons tagged eight to twelve newly hatched ducklings in each of over one hundred nests for an annual total of nearly one thousand ducklings. Later he could determine how large a percentage of tagged female ducklings returned as adults to nest in their own natal area. In total, his nestboxes have hatched over thirty thousand ducklings that, without his boxes, might never have seen the light of day. The data that Simmons has assembled have been cited in numerous references to the population and habits of Wood Ducks on the Pacific Flyway. Simmons also has built nearly two thousand Wood Duck nesthouses for sale at his cost to the public through the California Waterfowl Association.

Throughout the continent, over one hundred thousand nestboxes have been constructed, erected, monitored, and maintained by agencies and private individuals for their own pleasure and to benefit their favorite duck. In Minnesota, a group of friends of the Wood Duck creates enthusiasm and support for the species by publishing periodically the *Wood Duck Newsgram,* which is mailed to subscribers across the country. Elsewhere professional and amateur photographers sit for hours in blinds or paddle streams and bayous to take pictures of the beautiful Wood Duck. Over the years, hundreds of magazine

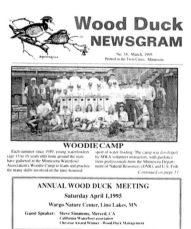

articles, photographs, sculptures, and calendars featuring Wood Ducks have educated and inspired conservationists, naturalists, hunters, and birders of all ages.

The national management effort on behalf of the Wood Duck has been unique and successful. The larger issues of the environment, which affect all wildlife and habitat, have been handled at the highest levels by the government and large environmental organizations, which possess the essential political clout and capital. These entities work, as Aldo Leopold has described, with the "steam shovel." Playing an equally important role are individual citizens with their "hand shovels," working to protect their favorite species.

THE MANDARIN OF EAST ASIA: A THREATENED SPECIES

The Wood Duck of North America has always been an important game bird. The rapid decline of its population was halted by state and federal legislation prohibiting hunting of the species for more than three decades. Unfortunately, such a simple solution is not available for the Mandarin. For many years, the species has been protected from legal hunting in all the Asian countries it inhabits. While accidental shooting and poaching are worrisome and continuing problems, the immediate threat to the Mandarin's existence in most of Asia is the relentless and increasing attack on its habitat—the forests where it finds nest cavities and food, and the waters and brooding areas where it raises its ducklings.

Environmental protection in the native lands of the Mandarin is for many reasons mostly beyond the influence of the individual. Russia is at a political crossroads, and economic needs may prevail, except in unusual circumstances, over environmental needs. Alarming damage is already being encountered in Ussuriland, where Russian, Korean, and American companies are logging off huge areas in prime Mandarin habitat. Mining, farming, and various industries are polluting once-pristine wetlands essential for Mandarin and their broods. In the face of such overwhelming problems, in 1989 Russia issued a national postage stamp featuring a painting by Ivan Koslov titled *Mandarin in Autumn.* Even Mongolia, where Mandarin have never been known to occur, has a special Mandarin stamp.

TOP Mongolia, a country without a population of Mandarin, joined the conservation effort on behalf of the species by issuing a postage stamp depicting the Mandarin drake.

ABOVE In 1989 Russia used a detail of Ivan Koslov's painting for a Russian stamp and sold an edition of prints based on the painting to spread awareness of the Mandarin and raise conservation funds.

OPPOSITE The character and beauty of the Mandarin are a convincing argument for preserving the species.

China has been in the throes of an economic explosion, which has resulted in the devastation of its forests and the pollution of its water and air. Ancient Mandarin habitat in central China has long been usurped by the country's burgeoning human population, and large-scale plans call for moving even larger numbers of people into those northeastern regions of China that have been the nesting areas of last resort for Mandarin. Destructive industrial and agricultural development is sure to follow. In both Russia and China, heroic efforts will be required to make conservation effective.

In Japan, conditions are more favorable for the Mandarin. Its winter population appears secure and may be increasing. The nesting and breeding potential, however, needs to be evaluated. With nesting habitat seriously threatened in Russia and China, it is of paramount importance to study Japan's nesting potential and take action to improve it by the wise use of nestboxes and the banding of birds both nesting and wintering. It is possible that, with proper planning and execu-

Industrial and commercial development invades the heart of Mandarin country—the once isolated and wild Bikin River valley in Ussuriland—seriously threatening nesting and brooding habitat.

tion, Japan could support, as a last resort, a self-sustaining population of Mandarin independent of the rest of the Asian flyway.

Private organizations, primarily in the United States and Great Britain, are assuming an active role in the conservation of wildlife worldwide, even including Mandarin territory in the remote area in southeastern Siberia known to the Russians as Ussuriland. A cooperative effort by Russians, Chinese, Japanese, Koreans, and Americans, led by Russian scientist Sergei Smirenski, has been working to preserve six species of cranes, four of them endangered, that inhabit the area. Smirenski was codirector of a symposium in July of 1992 known officially as the International Workshop on Cranes and

Storks of the Amur Basin. The conference was sponsored by the Socio-Ecological Union in Russia, and the National Audubon Society and the International Crane Foundation in the United States, with the support of other American organizations such as the Lake Baikal Watch, the Wilderness Society, and the Siberian Forest Protection Project. These last three organizations, already engaged in conservation efforts in the Lake Baikal region, joined the other organizations in their attempt to protect the Amur-Ussuri basin along the China-Russia border. Lake Baikal is the oldest and deepest lake in the world—holding one-fifth of the planet's supply of fresh water, more than all the water in the five Great Lakes of North America. Often referred to as the crown jewel of Russia's natural inheritance, the lake has attracted intense international cooperation to combat the industrial pollution that threatens it. This same international cooperation in the neighboring Amur-Ussuri river areas will directly benefit the Mandarin.

The immediate concern in the Amur basin is to prevent construction of the Khinganski dam, which, according to Sergei Smirenski, would flood vast areas of the basin and devastate the waterfowl habitat. Of equally urgent concern is the unchecked harvesting of old-growth timber throughout the area, mostly by Russian, American, and South Korean lumber companies. Amid the political and economic chaos, local officials, politicians, private citizens, and even the military are said to be selling every available asset, even those in designated reserves. Smirenski describes them as "looters." So valuable is the lumber in Ussuriland that Russia has given contracts to North Korean companies using forced prison labor to work in weather conditions that Russian laborers would not voluntarily endure.

The task of conserving this vast and wild area—one of the last of its kind in the northern hemisphere—is a challenge meriting international participation. With funding from the United States and Japan, the Socio-Ecological Union studied and then established a twenty-seven-thousand-acre wildlife sanctuary along the Amur River. As a direct result of this project, the group called for the Russian and Chinese governments to carry out a study on the environmental impact of the Khinganski dam on the basin. Other tentative accords were reached—on a potential international reserve at Lake Khanka at the Amur's southerly reaches above Vladivostok and a proposed survey of cranes, with Russia providing a helicopter and China providing logistical support for biologists from both countries. Noritaka Ichida, president of the Wild Bird Society of Japan, flew with a Russian scientist to the Kuril Islands to try to establish further cooperation to study a colony of cranes there. These small but critical first steps are the outcome of the interest and enthusiasm generated over the years by a handful of

individuals who founded organizations devoted to protecting species of crane and their habitat.

This international effort, primarily to benefit the cranes, concerns the conservation of habitat important as well to the Mandarin. As the crane has its supporters, so the Mandarin needs dedicated friends to organize on its behalf—in this critically threatened area.

For two quite unrelated reasons, the British may be the best friends the Mandarin has for protecting and conserving the species in Asia—nearly six thousand miles away. With Mandarin nesting habitat being destroyed rapidly in much of Ussuriland and northern China, it is only sound planning to envision a worst-case scenario whereby the Mandarin, like other species before it, is brought to the brink of extinction. For some species, the only saving resource was a gene pool, found most often in captive stock, whose offspring could be released into the same habitat that the species once occupied. Intro-

Migrating Bewick Swans have
wintered for many years at the
Wildfowl and Wetlands Trust in
England. The tails of some birds
are marked with yellow for possible
recognition on the species' Russian
breeding grounds.

ducing captive wildfowl into an area where a species has become extinct, or its
numbers severely diminished, is a hopeful but never certain solution. A half
century ago, the population of the Nene Goose in Hawaii had decreased to
fewer than fifty birds. Geese were reintroduced to their native habitat, and the
Nene population rose to five hundred, but warnings are again being heard
that unless feeding habitat, chiefly areas with native grasses, is restored, the
stock again will begin to decline and the Nene may disappear from the world
in the next fifty to sixty years.

Assisted by the release of captive birds into their natural habitats, the
Aleutian Canada Goose and the rare Brown Teal of New Zealand continue to
fight for their existence, often because the causes of their original decline have
not been corrected. But as Dr. Janet Kear writes in her award-winning book
Man and Wildfowl, many threatened wildfowl—as ducks, geese, and swans
are called in the United Kingdom—"adapt well to captivity and are being

bred in numbers that may ensure future survival, even after they have died out in their natural state."

Therefore, should the Mandarin be lost to the pollution and habitat destruction in the Far East, Britain's substantial stable population of healthy wild Mandarin offers not only a broad-based gene pool but, equally important, a supply of noncaptive, free-flying birds capable of surviving in the wild. The British Mandarin, brought directly from the wetlands of China in the early 1900s, were able to adapt to a foreign environment, and Mandarin from this stock could be expected to survive in Asia if future conditions warrant such a return.

The British also may offer the best hope for ensuring the survival of the Mandarin before such a desperate situation develops. Britain's interest in waterfowl throughout the world is legendary. Britain and Ireland are home to more species of breeding waterfowl than most countries in Europe. In winter, two and one-half million ducks, geese, and swans visit these island areas. So interested are the British in waterfowl that, according to wildlife historian Sir Christopher Lever, more waterfowl have been introduced to and naturalized in Britain than any other wildlife group. In Britain, in proportion to its size, more money, time, effort, and interest may be devoted to waterfowl and their wetland habitat than in any other country in the world.

Russian guide Andronich Chirapanov helped researchers from the Wildfowl and Wetlands Trust during their 1991 quest for the Scaly-sided Merganser in Mandarin habitat in Ussuriland.

At the height of the British Empire, it was a common pastime for British officials working in the colonies to explore unknown tracts of land and shoot unusual birds and mammals and send their skins home as curiosities or for study. Long after this exposure to unfamiliar lands, the British continue to find those areas of interest and even possibly feel a responsibility for them. Certainly their enthusiasm for wildfowl knows no geographical bounds.

Almost single-handedly, Sir Peter Scott attracted the far-ranging Bewick Swans, in their westward migration from their breeding grounds in Siberia, to the ponds at the Wildfowl and Wetlands Trust (WWT) at Slimbridge. There they were studied for so long and in such detail by the British that Dr. Paul Johnsgard was able to write that, as a result of these efforts, more is known today about this species than about any of the other northern swans. The British have taken the lead in working for the reestablishment of the Nene Goose in Hawaii—including the raising

of goslings for release into the wild—and have even provided most of the funding for research for this effort.

Another endangered species in which the WWT has
had a long-standing interest is the White-winged Wood
Duck, which lives on small streams, ponds, and swamps in
the rain forests of Southeast Asia, from Assam in northeast
India east to Vietnam and south to Sumatra in Indonesia.
As do other wood ducks, including the Australian Wood
Duck, the White-winged Wood Duck nests in hollows in
trees and has suffered drastically from deforestation. The
forested areas have been reduced by two-thirds, and the
remaining forests are being
lost at a rate of nearly two
percent each year. The first
White-winged Wood Ducks
to breed in captivity to the
second and third generations
did so at Slimbridge in 1969,
as the Nene Goose had done,
and the species has continued
to breed there ever since.

With the nesting habi-
tat of the White-winged Wood Duck declining rapidly, researchers decided to
search for populations throughout the species' range in the hope that the
forests where the ducks are found could be saved. In early 1993, the Wildfowl
and Wetlands Trust, working with the Royal Forest Department of Thailand,
launched such a search and in September located the largest population yet
discovered, at least seventy of these ducks, in four reserves near the Cambo-
dian border. Since then, more have been found in the forests of central Laos.

A similar effort of particular interest was undertaken in the summer of
1990 by the WWT and the International Waterfowl and Wetlands Research
Bureau (IWRB). The two organizations dispatched a senior field research
officer, Dr. David Bell, as the first British scientist to visit the forest wilderness
of the Russian Far East. There, in Ussuriland, he worked with Dr. Vladimir
Borcharnikov of the Far-East Branch of the U.S.S.R. Academy of Sciences.
Their assignment was to study and trap or net one of the world's rarest and
least understood wildfowl, the Scaly-sided, or Chinese, Merganser, which
nests in the upper reaches of the Bikin River, also the primary nesting area of
the Mandarin. The following summer, Dr. Barry Hughes, another WWT
researcher, with Dr. Borcharnikov and several other Russian scientists, built

TOP Vladimir Borcharnikov (right),
well known for his writings on
Russian wildlife, joined the 1991
expedition to Ussuriland to study
the Scaly-sided Merganser.
ABOVE LEFT Mandarin expert
Won Pyong-Oh of South Korea (left)
and scientist and photographer Yuri
Shibnev of Russia (sitting) stop
along the River Iman during the 1991
trip to Ussuriland.
ABOVE Shibnev's photograph of
the Bikin River flowing through the
vast forests of Ussuriland appeared
on the cover of the WWT magazine
that described his trip.

TOP One of the few photographs ever made of the rare Scaly-sided Merganser—a hen and her brood on the Bikin River—was taken by Barry Hughes, leader of the WWT expedition.

ABOVE On their next trip, WWT researchers hope to travel up the Bikin to study the Mandarin as well as the merganser.

on David Bell's experiences by continuing the study of the merganser in precisely the areas considered most important to the Mandarin during nesting season.

In typical British fashion, little publicity attended these efforts on behalf of the Nene Goose, White-winged Wood Duck, and Scaly-sided Merganser. *Wildfowl and Wetlands,* a color magazine published twice a year by the WWT, and a newsletter edited by the IWRB provided what little information was available on these and other endangered species. Under the heading "Taxa for Consideration," seventy-two waterfowl species were listed worldwide with classifications ranging from rare to possibly extinct. In 1992, the Mandarin was listed as rare in East Asia. Also listed as rare was the Marbled Teal, for which the IWRB and WWT had already developed and started to implement an International Conservation Plan. This suggested that if the teal were worthy of a conservation plan, the equally rare Mandarin might be a candidate for a similar plan. Christopher Savage, who had been an honorary consultant to the IWRB since 1965, was in England in the summer of 1994 and talked informally at Slimbridge with WWT and IWRB officials about the possibility of developing such a plan for the Mandarin. They expressed interest and suggested that I come to England to participate in this promising discussion with both groups.

At the first meeting on June 7, 1994, at WWT headquarters at Slimbridge, Savage and I were joined by Dr. Myrfyn Owen, director general of the trust, and Barry Hughes. The plan proposed by Owen and Hughes had nothing to do with the Marbled Teal. Instead, they intended to continue their work on the Scaly-sided Merganser in Ussuriland and would broaden it to include the Mandarin since both species are found in the same areas and in nearly identical habitat in Russia and China. The project had great appeal for two reasons. The Mandarin, introduced into Britain over sixty years ago, is held in great affection by the British, who understandably would wish to be active in their preservation. The second reason concerns funding for the project. Research on the Scaly-sided Merganser may be more difficult to finance as the merganser is not widely known outside scientific circles. Expanding the program to include the Mandarin would help immeasurably to attract financing for further research. It was a splendidly simple concept, a great tribute to the Mandarin, and a plan on which we could all agree.

At the meeting, Barry Hughes presented a report on his research on the Scaly-sided Merganser. As he related interesting details about his trip, and

BELOW Of all the regions in Asia, Ussuriland is believed to offer the best remaining forests and rivers for the Mandarin. Conserving Mandarin habitat is essential for conserving the species.

about the country and its people, he showed dozens of color slides that revealed Ussuriland to be a vast and forbidding wilderness—one that holds a special and, we hope, not fatal attraction for the Mandarin.

The Wildfowl and Wetlands Trust and other organizations at Slimbridge are involved in a rapidly increasing number of conservation efforts—sixty-five of the two hundred thirty-one species that the WWT describes as "wildfowl" are considered endangered. One project is of particular interest to those who have worked with the Mandarin. In 1989 another duck, the White-headed Duck, was classified by the International Union for the Conservation of Nature (now the World Conservation Union), or IUCN, as rare and possibly vulnerable (in danger of extinction in the near future). With this declaration, the White-headed Duck joined the Marbled Teal as the only European ducks threatened with global extinction. Its primary breeding habitat is in the former U.S.S.R., and its principal wintering habitat is on Lake Burdur in Turkey. Smaller colonies are found in Spain, Iran, Algeria, Pakistan, and Tunisia. The 1989 report by the IWRB notes that "the White-headed Duck is one of the rarest and most unusual wildfowl species in the Palearctic region" and concludes that the "most urgent priority for conservation action is extensive habitat conservation—and will require rapid and effective international coordination."

The conservation effort begun in 1989 by the WWT on behalf of the White-headed Duck focused almost exclusively on habitat preservation supplemented by severe hunting restrictions. By 1992, dramatic population increases were being reported. In Spain, for instance, there were believed to be only twenty to thirty birds in the late 1970s. By 1992, the winter count was almost eight hundred birds. Then, in 1993, came the frightening news that the chance introduction of an alien species of duck had the potential to do great and immediate harm to the White-headed Duck. The tragic irony is that this sudden threat, by all accounts, had its beginning almost fifty years earlier at the Wildfowl and Wetlands Trust in Slimbridge.

In the late 1940s, the trust imported three pairs of Ruddy Ducks from the United States as part of a study to learn more about their breeding biology. White-headed Ducks and Ruddy Ducks are members of the tribe known as Stiff-tailed Ducks. Even as late as 1978, the greatly respected international ornithologist Dr. Paul Johnsgard was writing that he did not believe the two ducks were closely related—a view that might have lessened any concerns for the problems developing but as yet unnoticed. The Ruddy Ducks at Slimbridge hatched their first ducklings in 1949. So precocial and independent are the ducklings that they often leave their mothers only a few days after hatching. Despite the best efforts of the keepers at Slimbridge, they were unable, as was their practice, to catch and pinion the fast-diving ducklings. A

Hybridization is always a potential threat to the integrity of a species. Mandarin and Wood Ducks, which have long been kept together in captivity, cross-mate only rarely. Like this Mandarin drake and Wood Duck hen, they never produce a hybrid.

few months later, the first free-flying Ruddy Ducks escaped from their open-topped pens, and by 1960, they first bred in the wild outside Slimbridge.

In January 1993, the WWT and IWRB released sobering news in a report that described the first sighting of the wandering Ruddy Duck in Spain in 1983 and of the first hybrid offspring of the Ruddy Duck and the White-headed Duck in 1990. It was noted that Ruddy Ducks and their hybrids are more aggressive than White-headed Ducks and that Ruddy males are dominant over White-headed males. Scientists have confirmed that, whereas many hybrid species are infertile, these hybrid males are fertile and the females are believed also to be fertile. Both are dominant over White-headed Ducks—a lethal combination. Ruddy Ducks have been sighted in eighteen countries including France, Belgium, the Netherlands, Iceland, Norway, Germany, Denmark, Switzerland, Spain, Italy, Finland, Portugal, Morocco, and possibly the Ukraine. Eighty percent of all White-heads breed in Russia, ten percent in Turkey, and small numbers in Iran, Algeria, Spain, and elsewhere in the region. Unless immediate remedial action is taken, the estimated twelve to fifteen thousand White-headed Ducks are facing almost certain extinction—literally being hybridized out of existence as a species.

The Wildfowl and Wetlands Trust was quick and professional in its response to this emergency in which they had a more than casual interest. In February 1992, the WWT convened a meeting of conservation organizations in the United Kingdom to consider what action should be taken. This led to an international workshop held in March 1993 at the WWT Centre at

TOP Released into Europe unintentionally, the aggressive Ruddy Duck of North America is spreading throughout the continent.
ABOVE The White-headed Duck—like the Ruddy Duck, one of the tribe of Stiff-tailed Ducks—was already classified as rare when it began to hybridize with the Ruddy Duck, further diminishing the population of purebred White-heads.

Arundel in England. The fifty-four representatives from ten countries concluded that urgent action was required to reverse the population growth and expanding range of the Ruddy Duck in order to safeguard the globally threatened White-headed Duck. The primary problem is in Great Britain, where some thirty-five hundred Ruddy Ducks breed and export their numbers to the continent and therefore need to be contained. Except in the Netherlands, where approximately two hundred birds have been sighted, relatively few have been found in the other eighteen countries. Yet wherever the Ruddy Duck is found in the territory of the White-headed Duck, the Ruddy must be eliminated. Hunting of the species must be encouraged at the same time that shooting of the White-headed Duck has to be prohibited to protect the remaining population. This presents another difficult problem. Since the two ducks have a distinctive and almost identical profile, there is obvious danger in advocating harvesting of one species at the risk of diminishing the other.

It is instructive to recall that nearly ten years earlier, while encouraging my study of the alien Mandarin in the United States, Starker Leopold had cautioned careful analysis of the risks involved in the introduction of exotic species. In England, where Ruddy Ducks have failed almost every test for desirable introduction, the Mandarin have done better. The introduced Mandarin in Britain and the United States are nonmigratory, but the Ruddy Ducks are lively migrants quick to invade distant territories. Mandarin are unable to hybridize with any other duck and are not an aggressive species. In the United States at least, their late nesting pattern forces them to forfeit their choice of prime nesting and brooding habitat to the indigenous Wood Duck—a nearly insurmountable handicap for Mandarin. Of immense importance, if other precautions to control the Mandarin were to fail, the appearance of the Mandarin drake clearly distinguishes it from other species of ducks in North America, thereby making hunting feasible as an alternative means to keep the population in check. In fact, if the Mandarin is not legally protected, the drake's unique sail feathers and coloring almost guarantee that hunters will take the birds as trophies wherever they are found.

The solution to the problem of the Ruddy Duck will require money, manpower, and an extensive network of cooperating organizations from the governmental level to the individual hunter. As Savage and I watched the conservation strategy evolve, it became clear that for rescuing a single species of wildfowl in distant lands, the Wildfowl and Wetlands Trust and its allies at Slimbridge—the International Waterfowl and Wetlands Research Bureau and its Threatened Waterfowl Research Group—were uniquely qualified and experienced. Under the umbrella of the Swiss-based World Conservation Union, the largest and perhaps the most effective and least understood

Waterbird's-Eye View of the World Conservation Union
1995

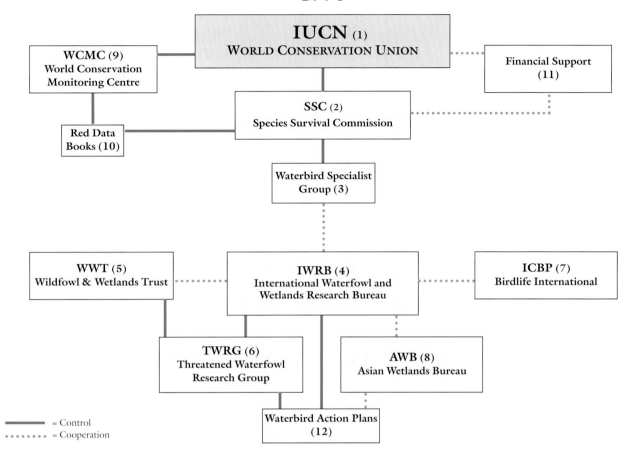

= Control

•••••••• = Cooperation

(1) **Worldwide Conservation Union (IUCN)** Founded in 1948 as the International Union for the Conservation of Nature and Natural Resources, and still known as the IUCN, it works to preserve and protect biodiversity and resources around the world. The IUCN is the only global conservation organization that brings together governments and their agencies and leading environmental organizations to articulate policy and implement specific projects. Members include most major conservation organizations, such as the World Wide Fund for Nature, Nature Conservancy, and National Audubon Society, and governments, corporations, and foundations. Headquartered in Gland, Switzerland.

(2) **Species Survival Commission (SSC)** One of six standing IUCN commissions, it is made up of a network of leading experts in the field. Over 5,000 volunteer scientists and field-workers provide advisory services and create Action Plans for the IUCN and its constituents. The world's largest professional network devoted to species conservation, the SSC makes its services available, often in partnership with other groups.

(3) **Waterbird Specialist Group** Managed by the IWRB, this one of 100 specialist groups is concerned with ducks, geese, and swans.

(4) **International Waterfowl and Wetlands Research Bureau (IWRB)** Founded in 1954, the IWRB is the only global nongovernmental organization devoted to the conservation of wetlands. It works with regional and international organizations to coordinate research projects and develop Action Plans to conserve wetland habitat. It also sponsors workshops and takes a leading role in the monitoring of waterfowl populations in 100 countries. Located in Slimbridge, England.

(5) **Wildfowl and Wetlands Trust (WWT)** Founded in 1946, the WWT maintains eight centers in the United Kingdom, where visitors come to view and learn about birds within their wetland habitats. A member of the IUCN, the WWT develops programs for globally threatened waterfowl in cooperation with the IWRB. Located in Slimbridge, England.

(6) **Threatened Waterfowl Research Group (TWRG)** A partnership of the WWT and the IWRB, this research group conducts in-depth studies of threatened species of waterfowl. Located in Slimbridge, England.

(7) **Birdlife International (ICBP)** This nongovernmental organization, which works closely with the IWRB, oversees population studies of all avian species. Located in Cambridge, England.

(8) **Asian Wetlands Bureau (AWB)** This organization works in collaboration with the IWRB and other agencies to inventory, study, and conserve waterfowl and wetlands in the Asia-Pacific region. Located in Kuala Lumpur, Malaysia.

(9) **World Conservation Monitoring Centre (WCMC)** Founded by the IUCN in the 1970s, it provides a worldwide information service based on its collection and analysis of global conservation data including the **Red Data Books** and the **Red Lists** of threatened species. Located in Cambridge, England.

(10) **Red Data Books** Started in the mid-1960s and supplemented in 1986 by the **Red Lists** and prepared jointly by the IUCN, WCMC, and SSC, these publications catalog the status of more than 5,000 threatened species of animals and plants and rate the degree of the threat.

(11) **Financial Support** The IUCN is funded primarily by member dues, including major contributions from the World Wide Fund for Nature, and from nearly 30 governments or their agencies, banks, foundations, and other sources.

(12) **Waterbird Action Plans** Prepared by the WWT and IWRB on behalf of the IUCN, SSC and Waterbird Specialist Group, these plans are the first step in implementing a major conservation effort on behalf of a threatened species.

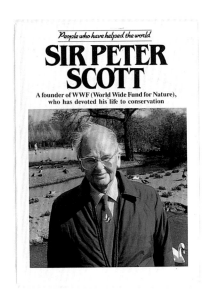

The late Sir Peter Scott's impact on global conservation can still be felt today. Cofounder of the World Wildlife Fund, he is the only individual ever knighted for his conservation work.

conservation organization in the world, these British organizations have access to a vast network of five thousand scientists and field-workers worldwide. The IUCN has been referred to as the nerve center of global conservation. Membership in the IUCN, or its Species Survival Commission (SSC), includes almost every major conservation organization in the world both private and governmental, among them the World Wide Fund for Nature (as the World Wildlife Fund is called in Europe), Nature Conservancy, National Audubon Society, National Geographic Society, Ducks Unlimited, Sierra Club, Wild Bird Society of Japan, and Wildfowl and Wetlands Trust. Officers of many of these organizations serve as volunteers on the SSC Steering Committee, so there is constant interaction among the IUCN, the SSC, and its worldwide membership.

The Species Survival Commission, one of six standing commissions of the IUCN, coordinates its efforts on behalf of waterfowl through the IWRB and its affiliate, the Threatened Waterfowl Research Group (TWRG), both headquartered at Slimbridge in England. There, the Wildfowl and Wetlands Trust provides both organizations with financing, office space, and personnel for special assignments, and publishes the IWRB-TWRG newsletter, which first highlighted the plight of the Marbled Teal and the White-headed Duck in Spain. The complexities of the IUCN, as one officer of the IUCN explained, cannot be described on an organization chart—the interrelationships are too fluid and change too rapidly as new challenges are met. The interactions could, however, be diagrammed but only at a single moment in time (see chart, page 209). He described how Peter Scott, who had chaired the Species Survival Commission for eighteen years—and before that had founded WWT—had left an indelible imprint on every aspect of its operations. Scott understood that endangered species throughout the world first must be identified before they can be saved, so he conceived what became known as the Red Data Books for all species of flora and fauna whose survival is threatened. He also recognized the importance of secure funding for such an ambitious undertaking. To meet this need, he helped create and became the first chairman of the World Wildlife Fund (WWF) as a primary source of funding for IUCN.

The net result of Scott's close relationship with the IUCN, the World Wildlife Fund, and the WWT—and of his personal friendship with the royal family in Britain and nobility throughout Europe, his effective fundraising in the United States, and his great success in bringing an esprit de corps to all those involved—was that his influence reached into the farthest corners of the globe. This is the IUCN network.

Large organizations such as the World Pheasant Association International in Britain and the International Crane Foundation in the United States focus attention and efforts worldwide to preserve these birds. Imposing animals such as elephants and whales will always find an outpouring of support for their protection. In most industrialized countries, other organizations are poised to help and are financed adequately to protect their native species. Should the Mandarin's close relation, the North American Wood Duck, ever again become an endangered species, Ducks Unlimited, the National Audubon Society, and other defenders would be fast, formidable, and effective allies.

But for threatened species like the Mandarin, indigenous to countries where people are often struggling to ensure their own survival, too few resources are available to expend much beyond providing the population with the most basic necessities. Endangered species such as the White-winged Wood Duck of Southeast Asia, the Marbled Teal, and even the struggling White-headed Duck fleeing its alien suitors can look to the concerned professionals at Slimbridge. They and their mostly unseen web of scientists and field-workers throughout the world may hold in their hands a safety net to protect these and other threatened species, including the Mandarin, which as yet have little other support.

Before conceding all responsibility for the Mandarin to the experts, it should be recalled that laypeople can also make a

A widespread nestbox and banding program, a top priority for conserving the Mandarin in Asia, would help provide nesting facilities where forests are being destroyed and supply essential information about population trends, migration routes, and availability of nesting habitat.

major difference. In Japan, for instance, where thousands of Mandarin spend the winter every year, as many as one-third might stay to nest if nest sites were available. There is reason to believe that central and northern Honshu suffers from an acute shortage of natural cavities. A program for installing man-made nestboxes is therefore an exciting and promising effort to begin. The recently established Oshidori Trust, inspired by Yuzo Murofushi, who has for years had his own nestbox program near Lake Ashi, could play a leading role.

Banding Mandarin—females caught on their nests or both females and males trapped during winter—should be a top priority to learn their nesting and migratory patterns. With expanded nesting facilities in Honshu and Hokkaido, it is possible to envision a self-sustaining Mandarin population in Japan, independent, if necessary, of the rest of the Asian flyway. In the United

States, professionals and wildlife enthusiasts provide nesthouses for Wood Ducks, band the ducks and study their behavior, and popularize the species. Similar programs in Japan could invite the participation of everyone from schoolchildren and members of youth organizations to retired businessmen.

In Japan, captive Mandarin increasingly can be on display in temple gardens and private ponds to familiarize people with the species. In China and Russia, similar programs to popularize the Mandarin are possible. As the remarkable recovery of the North American Wood Duck has proved, people working together can forge an alliance for helping the Mandarin, and other endangered species, which brings pleasure and pride to those who participate—and the world will be the better for their success.

[L.L.S.]

The exquisite grace of the Mandarin on this pond in Japan should inspire people in all the countries where the birds are native to cooperate in protecting the species.

A top-loading cedar nestbox at Indian Meadow Ranch, located thirty feet to the right of a natural cavity, has produced far more ducklings than the natural nest cavity.

NESTBOX PROGRAM FOR WOOD DUCKS OR MANDARIN

There are three somewhat different, but sometimes overlapping, reasons for participating in a nestbox program for Wood Ducks, Mandarin, or any other wild hole-nesting bird.

One is to provide potentially safe and expanded nesting facilities for those species whose nesting habitat is believed to be in decline. The first nestboxes for Wood Ducks were developed to offer nesting opportunities in areas where the scarcity of natural cavities was thought to limit the production of young. For many people, this is still the primary focus of a nestbox program for Wood Ducks.

Another reason is the pleasure of being involved with nature, the excitement of seeing the first birds search for and discover a nest site, and the ultimate thrill, in the case of Wood Ducks and Mandarin, of watching day-old ducklings follow their softly calling mother to her side in the water or on the ground below the nest.

Finally, there is the need for serious research, done mostly by professionals in the field, but also occasionally by those with a serious interest in nature and a desire to learn the nesting biology of specific species. In the United States, most research on migratory birds that involves invading a nesting in progress—banding hens, web-tagging ducklings, even the invasive handling of eggs or newly hatched birds—requires a federal and usually a state permit. Despite these restrictions, the nesting of Wood Ducks has been studied for many years by scores, perhaps hundreds, of professionals and amateurs. In the case of wild Mandarin in their native habitat, this type of research is just beginning. There is still much to be discovered, and nestboxes can play an important role in learning about the Mandarin's biology and ecology.

Over one hundred thousand nestboxes in active use throughout North America are estimated to add each year some one hundred fifty thousand wild Wood Ducks to the adult population. The California Waterfowl Association, a private nonprofit organization, has been working with the U.S. Fish and Wildlife Service and the California Fish and Game Department on a five-year project to increase the number of nestboxes in use in the state from a few thousand in 1990 to over seventy-five hundred by 1995. This program is typical of those in other areas across the continent. With the increasing interest in conservation and the ever-decreasing availability of many types of essential habitat, it seems inevitable and desirable that such nestbox programs for Wood Ducks and for Mandarin in the Far East be encouraged.

Constructing a nestbox is relatively easy and inexpensive. But a nestbox program, whether employing only a few or a hundred nestboxes, should not be undertaken without a full understanding of what is involved, for one's primary responsibility is for the enhancement of the production of young. Above all, the habitat for a nestbox program must be one in which the ducks can find adequate cover and a hen can be assured of adequate food for herself and her ducklings. Where those conditions exist, a program to provide greatly needed nesting facilities can be successful. Nestboxes must be well designed, well constructed, and carefully located to attract the ducks to a favorable environment. Twice each season, nestboxes should be checked, old debris removed, fresh shavings installed, and cracks, open knotholes, and any other physical damage repaired so that the nests are safe and attractive to the next year's prospective tenants when they arrive in the spring.

A well-designed and well-located nestbox may be occupied the first year it is installed, or only after several years—and sometimes never—depending on local population abundance. Patience and perseverance may be necessary to succeed in establishing a "colony" of nesting Wood Ducks. If ducklings are successfully hatched, hens may make a lifelong commitment to that nestbox or that area. Often some of the offspring will return to the area where they were hatched. To capitalize on the homing character of successfully breeding females, participants should continue and even expand the program for as long as possible.

Frederic Leopold and Arthur Hawkins were successful in attracting Wood Ducks to three nestboxes they installed in 1943 on Leopold's property on a 130-foot-high bluff overlooking the Mississippi River on the outskirts of Burlington, Iowa. Wood Ducks had nested nearby, so within a few years Leopold had more than twenty boxes in active use, which he and his family monitored until his death, at age ninety-four, in March of 1989. During most of the last fifty years of his life, he observed and recorded in detail every aspect of a female Wood Duck's routine, from the time she arrived to select a nestbox, through her laying and incubating of eggs, until approximately forty-five days later, when she led her ducklings down the bluff, across the railroad tracks, and onto the great river. Leopold's friends and neighbors were so fascinated by his project that many set out their own nestboxes. By 1965, he estimated that more than one hundred nestboxes had been erected in and around Burlington. Over twenty years later, a resident of Burlington remarked to Leopold that "it seems half the people in Burlington now have Wood Ducks nesting in their yards." An exaggeration, of course, but an indication that under the best of circumstances a few nesthouses can become the nucleus for a thriving breeding colony of Wood Ducks.

Nesthouses should be installed as near as possible to fresh water and never farther from the water than a mile. Typically they can be hung vertically on a tree trunk or post with the bottom of the box 3 to 4 feet above the ground to as high as 15 to 20 feet. The lower the box, the easier it is to install, inspect, and service, but the more likely it is to be disturbed by human activity or natural predators.

Three other factors are important in siting a nestbox. If possible, the entry hole should be clearly visible to the duck from the nearest water. Tree limbs, vines, or other vegetation that blocks easy flight to the nestbox must be removed each season. Finally, boxes should be so oriented that early morning sunlight does not shine directly into the nesthole. Hens tend to lay their eggs in the early morning, and the direct sunlight on the box suggests to them that the nest is not well hidden. When more secure cavities are available, hens also tend to avoid boxes with large cracks or loose knotholes that admit daylight. To a nesting hen, the darker the cavity interior, the better.

BUILDING A NESTBOX

Man-made nestboxes replicate the best features of a natural cavity in a tree. The cavities preferred by Wood Ducks and Mandarin have relatively small side entrance holes such as might be carved by a large woodpecker or a squirrel. Nestboxes must be large enough and deep enough to enable the nesting hen, her eggs, and ultimately her ducklings to be hidden mostly in darkness below the entry hole. The interior below the hole must be covered with hardware cloth, or the material used to make the nestbox must be horizontally grooved, to allow young ducklings to climb to the hole when they are ready to leave the nest. The bottom of a newly installed nestbox must be covered with 3 to 4 inches of clean pine shavings. Wood Ducks never bring nesting material to the nest and will not nest without it. Nesthouses are easy to fabricate from a variety of materials in ways that minimize the dangers from predators. To deter entry by Raccoons, the entry hole should be elliptical—3 inches high and 4 inches wide—and the center of the hole should not be less than 18 inches above the bottom of the box.

Generally made from wood, sheet metal, or plastic, nestboxes are of several basic designs: top, bottom, front, or side opening. Each material and each design have approximately equal appeal to a nest-searching hen and, with a few exceptions, serve a nesting hen equally, but serve the program manager in significantly different ways and should be selected accordingly.

Front

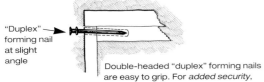

Top-lock "duplex" forming nails on both sides (see detail below)

5"

4"

3"

Wire mesh or hardware cloth stapled to *inside* of box below hole

12"

26" 18"

4"

3-4"-deep clean pine or fir shavings

Five 1/4"-dia drain holes drilled in bottom

3"

10–12"

Back of box with three holes, 1/8" dia, drilled for nail used to secure box to tree

Side

3/8"

1 1/2" overhang at front above hole

Cleats

Box on spike

Wire mesh

In back of box opposite entry hole drill 3/8" or wider hole to accommodate lag bolt or spike. Nestbox can be either *bolted* to tree (or post) for security or *hung* on a spike for ease of installation or removal. Bolt in a live tree should be backed off each year to avoid splitting box.

10–12"

To stabilize box, nail base securely to tree or post.

TOP-OPENING WOODEN NESTBOX

Materials
• 1" by 10" or 1" by 12" rough-cut cypress, cedar, or redwood
• 1" by 1" cypress, cedar, or redwood for cleats
• 4 "duplex" forming nails, #8, or #16 cut 2 1/2" long
• lag bolt, 1/4–3/8" dia by 4–6" long, with washer (or heavy nail or spike) for installing box
• wire mesh or hardware cloth, 4" wide (not screening)
• clean pine or fir shavings
• nails and screws, as needed, for assembling

Detail of Top Lock

"Duplex" forming nail at slight angle

Double-headed "duplex" forming nails are easy to grip. For *added security*, light or medium wire can be wound from nail head to nail head across top of box. Drill holes in ends of each cleat a minimum of 2" deep and 1/32" wider than forming nail so nail fits loosely and is easy to remove.

Underside of Top

Cleat located to fit inside box (including 1 1/2" overhang at front of box)

1" x 1" Cleats

5/8"

2"

Drill holes 2" minimum depth and 1/32" larger dia. than forming nail into end of cleats.

Cut cleats 1/4" shorter than inside width of box. Nail across grain of top.

Top-loading nestboxes are often difficult to access.

TOP-OPENING WOODEN NESTBOXES

This most commonly used of all styles of nestboxes is typically made of rot-resistant, rough-cut cypress, cedar, or redwood. *Advantages* Easily constructed from readily available and generally durable and low-maintenance materials, this style is as readily accepted by nesting ducks as other designs and materials. It is the best design for capturing a nesting hen for banding and other studies. Wood boxes appear more natural and therefore more attractive than those made of metal and plastic.

Disadvantages A wooden nestbox is relatively heavy—15–20 pounds. In areas where it is necessary to install boxes out of reach of human hands or inquisitive cattle, carrying a heavy box up a 20-foot ladder can be difficult or even hazardous. The ladder is required not only to hang the box, but also to inspect and refurbish it at least once or preferably twice a season. A 24-inch-deep top-opening box hung only 6 feet above the ground requires at least a small ladder to reach inside to the bottom to capture a hen, inspect her eggs, or clean the box. Wood is susceptible to cracking, loose knotholes, and damage by squirrels and woodpeckers, thus requiring greater maintenance than a sheet metal or plastic box.

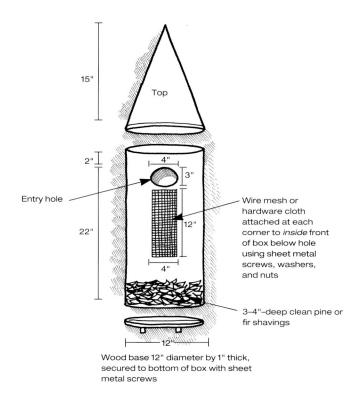

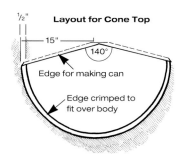

Layout for Cone Top

140°

Edge for making can

Edge crimped to fit over body

15"

1/2"

15"

Top

2"

Entry hole

4"

3"

22"

Wire mesh or hardware cloth attached at each corner to *inside* front of box below hole using sheet metal screws, washers, and nuts

12"

4"

3–4"–deep clean pine or fir shavings

12"

Wood base 12" diameter by 1" thick, secured to bottom of box with sheet metal screws

---

Underside of Base

1" by 1" cleats

1" by 6" boards cut across grain and joined to make 12"-dia base, drilled with four or five 1/4" drainage holes

seam

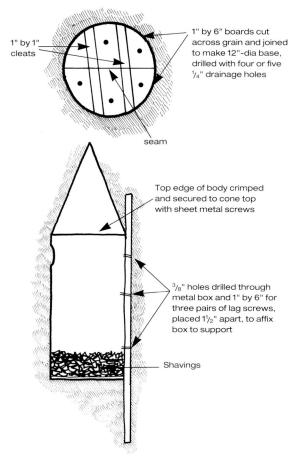

Top edge of body crimped and secured to cone top with sheet metal screws

3/8" holes drilled through metal box and 1" by 6" for three pairs of lag screws, placed 1 1/2" apart, to affix box to support

Shavings

TOP-OPENING SHEET METAL NESTBOX

Materials
- 26-gauge galvanized sheet metal (cold-air pipe)
- 1" by 6" rough-cut cypress, cedar, or redwood for base of box and back support
- Wire mesh or hardware cloth (not screen)
- Sheet metal screws, washers, and nuts, as needed, for fabricating box and cone, for attaching wire mesh or hardware cloth
- Wood screws as needed for making base
- 6 lag screws, 5/16" dia by 4–6" long, for attaching box to 1" by 6" support
- clean pine or fir shavings

TOP-OPENING SHEET METAL NESTBOXES

This box is nearly identical in its functional design to the top-opening wooden box.

Advantages Considerably more predator-proof than either wood or plastic boxes, this top-opening box is less than half the weight of a wooden box and requires almost no maintenance. It serves equally well for catching a nesting hen for banding and other research.

Disadvantages Sheet metal boxes are more difficult and costly for a layman to fabricate. The material is not as attractive initially to nesting hens or as aesthetically pleasing to humans. Like a top-opening wooden box, a sheet metal box requires a ladder for installation and inspection.

Because of ease and economy of construction and more natural appearance, a top-opening wood box is generally preferred over sheet metal except in the Deep South, where small Raccoons and certain snakes make metal boxes more practical. According to Frank Bellrose, these two styles of boxes account for eighty-five percent of the nesthouses in use in North America.

Great success is also reported using 8-inch-diameter and 10-inch-diameter PVC pipe for the body of a top-opening nesthouse and wood for the top and bottom.

BOTTOM-OPENING PLASTIC NESTBOXES

Over twenty years ago at Indian Meadow Ranch in northern California, a bottom-opening box constructed from two 5-gallon plastic paint or oil buckets was developed to facilitate research that required servicing a large number of boxes. Upon quick release, the bottom half of the nest can be removed, and the nest can be cleaned and refurbished in just a few minutes. More important for research is that a bottom-opening box facilitates a number of tasks: counting, weighing, and calibrating eggs, weighing day-old ducklings, analyzing and photographing in full daylight ducks and even competitors like bees and newly born squirrels and screech owls.

Advantages When used buckets are available, from a painter, for example, this design is very inexpensive to fabricate. Clean buckets can often be obtained at substantially less than the cost of materials for a wooden or metal box. Used buckets are easily cleaned with a wire brush, and with a little practice the entire nestbox can be built in fifteen minutes or less with only a

Bottom-loading bucket boxes are held together by two expansion springs or wire that can be released for quick inspection.

1/4-inch drill, a keyhole saw, a piece of hardware cloth, and some baling wire or the equivalent. Extremely lightweight—less than 5 pounds—these boxes require very little maintenance. The exterior needs only an occasional touch-up with light-colored paint. These plastic boxes are readily accepted by nesting Wood Ducks in direct competition with wooden nestboxes. Squirrels generally avoid them, perhaps because plastic is noisier than wood.

Disadvantages Plastic boxes, even when properly painted, are not as aesthetically pleasing as wood boxes. The life expectancy of a plastic bucket (.090-inch wall thickness) is only seven to ten years, although many of those in shaded areas at Indian Meadow Ranch in northern California are still sound after nearly twenty years. Plastic, more so than metal or wood, is subject to internal heat buildup if exposed to direct sunlight for long periods and can cause hens to desert or embryos to die, though this has not been a problem in northern California.

The Tom Tubbs is an excellent box with threads molded into the top and bottom sections.

The best bottom-opening nestbox for Indian Meadow Ranch was a Tom Tubbs brand plastic box made for commercial sale in the Minneapolis, Minnesota, area. In principle, the Tom Tubbs box is similar to the plastic bucket box. In practice, it was more durable, more aesthetically pleasing, and easier to operate and maintain, and for a product sold commercially it was reasonably priced. Like the plastic bucket box, the Tom Tubbs box was reported to experience an unacceptable heat buildup when exposed to direct sunlight for prolonged periods of time. Unfortunate publicity on this problem from some states in the Deep South may have been the reason that the sale of Tom Tubbs boxes was discontinued in the late 1980s.

In the middle 1980s, nearly two dozen of these boxes were installed at Indian Meadow Ranch, and they are still in use. No adverse effects from heat have been identified—temperatures in northern California from March through June are relatively mild, and most of the boxes are in shaded areas. Favorable mention is made of the Tom Tubbs style of box only to encourage the production of a similar bottom-opening box, i.e., attractive, durable, easy and economical to service, and exceptionally well suited to certain types of research that are not readily accomplished in a typical top-opening box.

In 1993 the Minnesota Waterfowl Association began trial-testing a unique design of bottom-opening box. To permit fast and economical servicing and inspecting of nestboxes that are installed high enough to escape curious visitors or vandals, these boxes have a deep molded plastic nest cup that fits securely inside the open bottom of this otherwise all-wooden box. At installation, the plastic cup can be dressed with fresh shavings, quick-

This unique bottom-loading box, distributed by the Minnesota Waterfowl Association, has great potential for saving time servicing nestboxes high off the ground.

locked to an adaptor on a 6- to 8-foot-long pole, and inserted into place. A twist of the pole locks in the nest cup, and a reverse twist releases the pole from the cup. At this stage of their development, these boxes are enthusiatically accepted by those who manage large numbers of boxes, where time and labor costs are particularly important. Certain design features for securing the bottom still remain to be perfected, but the concept is excellent.

FRONT-OPENING PLASTIC NESTBOXES

Introduced in 1988 and first publicized by Ducks Unlimited, this box is notable for its double-walled, heat-resistant construction. In appearance, it is equally as attractive as the Tom Tubbs. Although this front-opening box is thoroughly adequate for most purposes, like a bottom-opening box it is not suitable for capturing a setting hen on the nest. Its removable plastic tray, while handy for most research purposes, is not quite deep enough nor large enough to be as workable as the bucket box or the Tom Tubbs box. Despite these limited reservations, it is an excellent nestbox for most applications.

This front-opening nestbox is well insulated and easy to use.

TOP-OPENING WOOD NESTBOXES WITH SIDE ACCESS DOOR

Made to the same specifications as the basic top-opening wooden nestbox, these boxes have been developed to facilitate research. A 6-inch-diameter hole cut 4 inches above the bottom of the box is covered by a 6-inch-diameter plywood door hinged at the bottom. Eggs and chicks can conveniently be removed through the side door for study. Most of the fifty or more top-opening wooden nestboxes at Indian Meadow Ranch have been successfully modified to include the small side-opening door.

A top-loading box is modified with a side door, hinged at the bottom, for access from ground level.

Advantages This multiple-purpose box offers convenient access through the top for safely catching and banding a nesting hen. From the side-opening door, the nest can be observed, the eggs and duck-lings can readily be removed, and the membranes from hatched eggs can be sifted and counted. Removing old shavings and replacing them with fresh shavings are accomplished without having to work from the top in a darkened interior occupied perhaps by bees, snakes, or newly born squirrels.

Disadvantages Cutting a 6-inch-diameter side-entry hole and constructing a light-proof, quick-entry door add to the time and cost of a conventional wooden box. Like any top-opening wooden box, this style is heavy, but side-entry access is more readily available without the use of a ladder, except when catching a nesting hen. For photographing eggs or day-old ducklings, it is not as useful as the bottom-opening style.

PREDATOR DETERRENCE

In some instances all nestboxes, even those made of sheet metal, benefit from additional precautions against predation. There is, however, no way to ensure protection from hawks and owls. Some enter the nest to kill the hen; others take the hen as she leaves or enters the nesthole.

1. Nestbox entry holes should be checked annually and maintained at the minimum recommended shape and size. Squirrels will frequently enlarge the holes, and holes on some commercially sold wooden or plastic nestboxes may be too large or too round for protection again Raccoons.

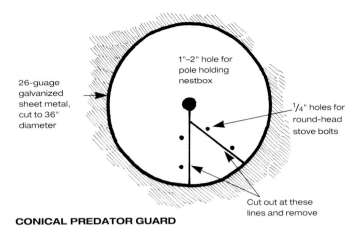

26-guage galvanized sheet metal, cut to 36" diameter

1"–2" hole for pole holding nestbox

1/4" holes for round-head stove bolts

Cut out at these lines and remove

CONICAL PREDATOR GUARD

2. For added protection, nestboxes can be mounted on metal poles sunk firmly into the bottom of a pond or stream, allowing sufficient height above high-water levels. Raccoons and some snakes swim well, however, and may not be completely deterred. A conical predator guard, made from sheet metal (above), is a practical deterrent on post-mounted nestboxes.

3. For tree-mounted boxes, the entire girth of the tree may be encircled by a piece of smooth, undented, unpunctured light-weight sheet metal at least 3 feet long, placed with the bottom at least 4 feet above the ground. Adjustments are required as the girth of the tree enlarges. The tree must be sufficiently isolated from the boughs of other trees that Ringtails, Bobcats, Raccoons, and squirrels cannot leap to the nest tree from adjoining unprotected trees.

4. Perhaps the greatest protection of all is a nestbox located near human habitation where predators fear to venture. Wood Ducks and Mandarin are more easily acclimated to civilization than many predator species.

Tips for a Successful Nestbox Program

1. In addition to good nesting facilities, the area must include good cover and good brood habitat—food, shelter, and areas for undisturbed rest. Be patient. Select the proper nest design, and install the boxes in what you believe to be an appropriate nest location. Scattering grain nearby on a regular basis often helps to attract ducks. If nests are not used immediately, let a year or two go by before moving them. Nests that have been unused for several years should probably be relocated and reoriented. For unknown reasons, some nests in what seem perfect locations are never occupied, even when other nests in the area are in almost constant use. Nest location is largely a matter of trial and error.

2. Nests should hang solidly on the tree. On a tree that is not entirely vertical, the top or bottom of the nestbox can be shimmed so that it is properly oriented. Branches that would impede easy flight to the nest hole must be removed. Accessibility of the hole should be checked annually.

3. In early February (or thirty days before first nest searches are expected), nests should be carefully inspected; knotholes, cracks, and ladders inside the boxes below the entry hole repaired; old shavings removed; and fresh shavings or sawdust installed to a depth of not less than 3 inches or more than 4 inches.

Stephen Simmons, based on long years of experience in central California, emphasizes the importance of constant supervision of nestboxes. He estimates that if one hundred boxes were installed and never inspected, cleaned, and repaired, none of the boxes would be usable by the end of ten seasons, due to theft, damage from storms, fallen trees, or nest competitors, blockage of entry holes by branches, or soiling and spoiling of the nesting material.

4. When forty to fifty percent of the nestboxes are used, more boxes should be installed.

5. If Wood Ducks or Mandarin are not attracted to the area, ducklings four to six weeks old can be imported. Keep them penned in the area until they are nearly ready to fly, then release them. Hens, with their well-developed homing ability, will frequently return to the area where as young ducklings they were held and released. Several spring releases, involving twelve or more ducklings, are desirable. Likewise, eggs may be imported, incubated, and hatched, and the ducklings released. It is important to check with local authorities to learn what prohibitions there may be against such transfers.

6. Each year between July and September, when the eggs have hatched and the ducklings have left the nest, nestboxes should be checked for use by Wood Ducks or Mandarin, the egg membranes counted, and other pertinent information noted in permanent record books.

7. When inspecting any nest at any time, unless a hen is to be captured, a hen on the nest should be allowed to escape as normally as possible. It is best to inspect nests in the early morning or in the late afternoon, when the hen would normally be off the nest feeding and resting, or would be preparing to leave the nest. To avoid unduly frightening a hen on the nest, it is good practice to scratch or tap on the tree to create the effect of a natural predator. Of nesting hens inspected with reasonable care, only about three out of every hundred will desert the nest, and the closer the eggs are to hatching, the less likely she will desert.

8. Finally, make sure that every action you take is *primarily* for the good of the species and only secondarily for your own pleasure. Unless research is a primary objective of the program, the less the birds are disturbed, the more successful the program should be.

BANDING MANDARIN IN THE FAR EAST

Banding waterfowl to trace their patterns of migration or to assist in efforts to project species population is practiced in many countries with varying degrees of sophistication. Wood Ducks in North America are banded as part of an ongoing program for migratory game birds supervised by the Bird Banding Laboratory of the U.S. Fish and Wildlife Service (USFWS). Band recoveries, mostly from birds brought down by hunters during the fall hunting season, provide information that is used to measure survival rates and homing and migrational patterns. When supplemented by other data, band recoveries help infer population estimates.

Official leg bands are provided to authorized banders by the USFWS and are numbered so that they may be matched with the bander's report filed with the USFWS showing species, age, sex, and date the bird was banded. All recovered bands are supposed to be sent to the Bird Banding Laboratory. However, only about one-third of the bands taken are actually returned. Issuing rewards for bands returned has increased the percentage of returns. The use of bands other than those issued by the USFWS is discouraged.

In the Far East today, Japan leads the way with banding programs coordinated with the Wild Bird Society of Japan (see Chapter Four). Such programs are concerned primarily with game birds and do not include the officially protected Mandarin. In the interest of conserving the Mandarin, others in the Far East may wish to establish their own banding programs and develop communications within their ranks in the hope of encouraging more formal programs.

In the spring, female Mandarin can be captured for banding while incubating eggs inside the nestbox. The *top-opening* nestboxes shown in Appendix I serve well as "traps" for nesting birds in spring. When preparing to remove a hen from her nest, it is advisable to block the nesthole to darken the nest and cut off the hen's escape route. A simple handmade device—a 4-inch-square piece of plywood with a hook to hold it in and against the hole, attached to a 6-foot-long handle—can be quickly and quietly installed by one person before climbing the ladder. From a darkened nest, a hen can usually be removed from her eggs by hand with a minimum of disturbance, banded, and gently placed back on her eggs. Or, after banding, females can be released and will normally return within an hour or so.

While the hen is off the nest, newly hatched ducklings, less

In preparation for catching and banding a nesting Mandarin hen, the entry hole has been blocked to darken and secure the nest before she is lifted from her eggs.

than twenty-four hours old, can be picked up in much the same way and web-tagged through a web between their toes. These tags are specially designed for the fin-tagging of fish and must be obtained privately. Banding and web-tagging of ducks and ducklings are useful in determining the increase or decrease in the number of hens returning to nest the following spring. Up to ninety percent of hens that have successfully hatched ducklings will return to nest in the same nest or in the immediate area, if they survive the wintering months. If the number of banded hens returning to a wide enough area for a long enough period of time is more than fifty percent of the total number of nesting hens banded, the species can be assumed to be on the increase. If less than fifty percent of banded hens are returning, the species is on the decline, and reasons for the apparently increased mortality must be identified. Of the fifty thousand to one hundred thousand Wood Ducks banded each year in the United States and Canada, fewer than one percent are banded in the nest and the rest are caught in so-called box traps of various designs (see page 171).

Box traps are used to capture adult males and females and fledglings in late summer and fall prior to hunting season. Traps in varying shapes and sizes are fabricated with chicken wire or small-mesh aviary wire over wood frames. Box traps are baited with grain to entice ducks to enter them. Some are large enough to accommodate several people so

This Y-trap is ready to be assembled and baited. The cover must be placed over the atrium, and the wings must be clipped into place.

that they can net and band captured birds. Other traps are small enough to be carried by hand. Six to ten can fit into a pickup truck for transportation to potential trapping areas. One smaller trap, known in the profession as the Y-trap for its unusual design, is of special interest. Easy and economical to fabricate, it is highly effective in catching and holding up to forty birds at a time. Of equal importance, birds can easily be removed for banding with minimum stress to the birds. This is an ideal trap for amateurs.

Box traps must be carefully supervised and regularly inspected. In summer and fall, traps should not be set out until all hens are off the nest and all ducklings in the area are large enough to survive on their own. Wood Ducks and Mandarin feed on land more often than most other ducks, so traps can be placed on land near the water, above the high-water level, or in the water on rafts or floats.

Traps must be checked regularly before midday every morning to catch, band, and release birds captured during the morning feeding hours and again just before dark to catch the evening feeders. Songbirds entering the trap for grain can be a problem. One remedy is to cut in the upper end of each wing an opening that is large enough to let a songbird escape and not so small that a duck will get its head caught. All traps should be dismantled and removed from the field when the program is ended.

As with most other box traps, birds are enticed to enter through tunnel-shaped entry corridors. Each corridor is large

enough at the entry end to permit easy access for a hungry duck and is small enough at the exit end, which extends approximately 6 inches inside the trap, to discourage the duck from attempting to escape. Birds trapped in a wire enclosure tend to run back and forth, probing the wire sides for holes large enough to squeeze through. Thus they dart around the small end of the corridor, never recognizing it as an escape hole.

The key feature of the Y-trap is the centrally located feed station, or atrium, into which the birds are guided by the three wings or holding pens. Having been attracted by the grain to enter the atrium through the tunnels, they realize that they are trapped and look for escape holes, which are the large openings of the tunnels leading into the wings.

The smaller the trap, the more likely that a trapped bird will eventually find and escape through the small end of one of the tunnels. The beauty of the small Y-trap design is apparent when an occasional duck is seen to leave one holding pen or wing only to find itself back in the atrium, from which it can easily pass through the large end of a tunnel leading into yet another wing. So effective is the design that no bird ever remains inside the atrium for more than a few moments before finding itself in another wing of the trap.

To remove and band trapped birds, each of the three wings can be detached from the atrium. Before a wing is removed, its open end should be covered with a board or fabric,

LEFT A Y-trap fully assembled—atrium covered, three wings attached—is placed atop an eight-foot square floating raft. These Wood Ducks and Mandarin were trapped at Indian Meadow Ranch during their early morning feed.
BELOW The Y-trap wings are sized so that birds are subjected to minimum stress when caught and removed for banding.

and the open side of the atrium closed in the same way. Every bird in the wing can be reached and quickly and gently caught by hand, banded, and released only slightly ruffled.

Because the atrium is small, the bait must be restricted in its distribution so that a duck must pass completely through the tunnel to reach the grain. If the grain is too close to the end of the tunnel, the bird can reach it while standing inside the tunnel and then can simply back out and escape. Unlike larger box traps that are commonly made of aviary wire atop a wood frame, the Y-trap is best made from more rigid rabbit wire or hardware cloth with ¼-inch-square mesh. When used in these dimensions, the material is self-supporting without the need for a frame. Tunnels are typically 8 to 9 inches wide at the large end, and 6 inches in length, with the small end 4 to 5 inches in diameter, only slightly larger than the entry to a nestbox. Tunnels can be oval or round, but are most easily fabricated square or rectangular.

Mandarin, in their native habitat in Russia, China, South Korea, and Japan, are not legally hunted, and because they are not considered game birds, they are not part of any official banding program. Although an effective waterfowl banding program normally depends on band recoveries by hunters, bands can also be recovered from birds struck down by natural causes, or from birds caught on the nest, where successful nestbox programs are in progress. Other band recoveries will result if a successful box-trapping program is undertaken in the Mandarin's nesting areas in late summer and fall and certainly on their wintering grounds in southern China, South Korea, and Japan. For example, hens banded in the spring on their nests in Russia may later be trapped and identified in Japan. Those same banded hens may be found on their nests back in Russia the following spring. Or a hen box-trapped and banded in the winter in South Korea may be found nesting in northern China. A banded drake may be shot by a poacher in Japan, and the band recovered and reported. The opportunities are endless and exciting, and the knowledge gained can be of tremendous value in the effort to learn more about conserving the Mandarin in its native lands.

RESOURCES

BIBLIOGRAPHY

The books, pamphlets, and articles listed below are among the resources that provide valuable information on the Wood Duck and the Mandarin.

Audubon, John James. THE BIRDS OF AMERICA. New York: Macmillan, 1937.

Bellrose, Frank C. DUCKS, GEESE AND SWANS OF NORTH AMERICA. Harrisburg, Pa.: Stackpole, 1942.

————. HOUSING FOR WOOD DUCKS. Urbana, Ill.: Illinois Natural History Survey Circular 45, 1953.

Bellrose, Frank C., and Daniel J. Holm. ECOLOGY AND MANAGEMENT OF THE WOOD DUCK. Mechanicsburg, Pa.: Stackpole, 1994.

Bent, Arthur Cleveland. LIFE HISTORIES OF NORTH AMERICAN WILD FOWL. Vol. 2. London: Dover, 1962.

Davies, Andy. "The Distribution and Status of Mandarin in Britain." BIRD STUDY 35 (1988): 203–208.

Delacour, Jean. "The Family Anatidae." WILSON BULLETIN 57 (1945): 26–31.

————. THE WATERFOWL OF THE WORLD. Vol. 3. London: Country Life, 1954–64.

————. THE LIVING AIR: MEMOIRS OF AN ORNITHOLOGIST. London: Country Life, 1966.

Dixon, Joseph. "Nesting of The Wood Ducks in California." CONDOR 26 (1924): 41–46.

Foster, Laura Louise. KEER-LOO: THE TRUE STORY OF A YOUNG WOOD DUCK. Healdsburg, Calif.: Naturegraph, 1965.

Fredrickson, Leigh H., et al., eds. THE 1988 NORTH AMERICAN WOOD DUCK SYMPOSIUM. St. Louis, Mo.: North American Wood Duck Symposium, 1990.

Fujimaki, Yuzo, and Yuzo Murofushi. DISTRIBUTION AND ABUNDANCE OF THE MANDARIN DUCK IN JAPAN. N.p., 1985.

Graham, Frank. "U.S. and Soviet Environmentalists Join Forces Across the Bering Strait," AUDUBON 93 (1991): 42–60.

Grey, Viscount of Fallodon. THE FALLODON PAPERS. Boston: Houghton Mufflin, 1926.

————. THE CHARM OF BIRDS. New York: F. A. Stokes, 1927.

Grice, David, and John P. Rogers. THE WOOD DUCK IN MASSACHUSETTS. Boston: Massachusetts Division of Fish and Game, 1965.

Grinnell, Joseph, Harold C. Bryant, and Tracy I. Storer. "The Wood Duck in California." CALIFORNIA FISH AND GAME 1 (1915):1–14.

A GUIDE TO INSTALLING AND MANAGING WOOD DUCK BOXES. N.p.: California Waterfowl Association, 1994.

Hawkins, Arthur S., et al., eds. FLYWAYS. Washington, D.C.: U.S. Department of the Interior, Fish and Wildlife Service, 1984.

Hester, F. Eugene, and Jack Dermid. THE WORLD OF THE WOOD DUCK. Philadelphia: J. B. Lippincott, 1973.

Hu, Hongxing, and Cui Yubo. THE EFFECT OF HABITAT DESTRUCTION ON THE WATERFOWL OF LAKES IN THE YANGTZE AND THE HAN RIVER BASINS. N.p., 1989.

Ji, Zhao, Zheng Guangmei, Wang Huadong, and Xu Jialin. THE NATURAL HISTORY OF CHINA. New York: McGraw Hill, 1990.

Johnsgard, Paul A. DUCKS, GEESE AND SWANS OF THE WORLD. Lincoln, Nebr.: University of Nebraska Press, 1978.

Kallman, Harmon, et al., eds. RESTORING AMERICA'S WILDLIFE: THE FIRST 50 YEARS OF THE FEDERAL AID IN WILDLIFE RESTORATION (PITTMAN-ROBERTSON, 1987). Washington, D.C.: U.S. Department of the Interior, Fish and Wildlife Service, 1987.

Kear, Janet. MAN AND WILDFOWL. London: T. & A. D. Poyser, 1990.

Knystautas, Algirdas J. V. THE NATURAL HISTORY OF THE U.S.S.R. New York: McGraw Hill, 1987.

Knystautas, Algirdas J. V., and Jurij B. Shibnev. DIE VOGELVELT DE USSURIENS (THE BIRDS OF USSURILAND). Hamburg: Paul Parey, 1987.

Laycock, George. THE ALIEN ANIMALS: THE STORY OF IMPORTED WILDLIFE. New York: Natural History Press, 1966.

Lee, Elizabeth. "Sharing the Earth." AUDUBON 95 (1993), 106–111.

LeMaster, Richard. THE GREAT GALLERY OF DUCKS AND OTHER WATERFOWL. Chicago: Contemporary Books, 1985.

Leopold, Aldo. A SAND COUNTY ALMANAC. London: Oxford University Press, 1949.

Leopold, Frederic. "A Study of Nesting Wood Ducks in Iowa," CONDOR 53 (1951): 209–220.

Lever, Christopher. NATURALIZED ANIMALS OF THE BRITISH ISLES. London: Hutchinson, 1977.

———. NATURALIZED BIRDS OF THE WORLD. London: Harlow, Longman, 1987.

———. THE MANDARIN DUCK. Princes Risborough, England: Shire, 1989.

Linduska, Joseph P., and Arnold L. Nelson, eds. WATERFOWL TOMORROW. Washington, D.C.: U.S. Department of the Interior, Fish and Wildlife Service, 1964.

Litvinenko, N. M., ed. RARE AND ENDANGERED BIRDS OF THE FAR EAST. Vladivostok: U.S.S.R. Academy of Sciences, Far East Branch, 1985.

———. RARE BIRDS OF THE FAR EAST AND THEIR PROTECTION. Vladivostok: U.S.S.R. Academy of Sciences, Far East Branch, 1988.

Lu, Jianjian. WETLANDS IN CHINA. Shanghai: Normal University Press, 1990.

Matthiessen, Peter. "The Last Cranes of Siberia." NEW YORKER 69 (1993), 76–86.

Miller, W. de W. "The Secondary Remigies and Coverts in the Mandarin and Wood Ducks." AUK 42 (1975): 41–50.

Palmer, Ralph, ed. HANDBOOK OF NORTH AMERICAN BIRDS. Vol. 3. New Haven and London: Yale University Press, 1976.

Phillips, John C. A NATURAL HISTORY OF THE DUCKS. Vol. 3. Boston: Houghton Mifflin, 1925.

Ripley, Dillon. A PADDLING OF DUCKS. Washington, D.C.: Smithsonian Institution Press, 1957.

Savage, Christopher. THE MANDARIN DUCK. London: A. & C. Black, 1952.

Scott, Derek A. A DIRECTORY OF ASIAN WETLANDS. N.p.: International Union of Concerned Scientists, 1989.

Todd, Frank S. WATERFOWL: DUCKS, GEESE AND SWANS OF THE WORLD. San Diego, Calif.: Sea World Press, 1979.

Trefethen, James B., et al., eds. WOOD DUCK MANAGEMENT AND RESEARCH: A SYMPOSIUM. Washington, D.C.: Wildlife Management Institute, 1966.

Wesley, David E., and William G. Leitch, eds. FIRESIDE WATERFOWLER. Harrisburg: Pa.: Stackpole, 1987.

Xiyang, Tang. LIVING TREASURES: AN ODYSSEY THROUGH CHINA'S EXTRAORDINARY NATURE RESERVES. New York: Bantam, 1978.

ORGANIZATIONS

The organizations listed below are among the many that, directly or indirectly, contribute to the welfare of waterfowl throughout the world.

Ducks Unlimited
1 Waterfowl Way
Memphis, TN 38120
U.S.A.

Ducks Unlimited Canada
Oak Hammock Marsh Conservation Center
Stonewall P.O. Box 1160
Oak Hammock Marsh, Manitoba
ROC 2ZO
Canada

International Waterfowl and Wetlands
Research Bureau
Slimbridge, Gloucester
GL2 7BX
England

National Audubon Society
700 Broadway
New York, New York 10003
U.S.A.

National Wildlife Federation
1400 Sixteenth St. NW
Washington, D.C. 20036
U.S.A.

The Nature Conservancy
1815 North Lynn St.
Arlington, Virginia 22209
U.S.A.

North American Wildlife Foundation
102 Wilmont Rd., ste. 410
Deerfield, Illinois 60015
U.S.A.

Wild Bird Society of Japan
Higashi 2-24-5
Shibuya-Ku
Tokyo 150
Japan

Wildfowl and Wetlands Trust
Slimbridge, Gloucester
GL2 7BT
England

Wildlife Management Institute
1101 Fourteenth St. NW, ste. 725
Washington, D.C. 20005
U.S.A.

World Conservation Union
Rue Mauverney 28
CH-1196
Gland
Switzerland

World Wide Fund for Nature
Avenue du Mont-Blanc
CH-1196
Gland
Switzerland

World Wildlife Fund
1250 Twenty-fourth St. NW, ste. 500
Washington, D.C. 20037
U.S.A.

INDEX

Adaptation
   evolutionary, 94–95
   to man-made nestboxes, 24, 45, 48,
      76–80, 84, 98–99
   of migratory instinct, 18, 48–49, 164–165
   See also Habitat; Nest sites
Adoption, 103–107, 112
Aix, 9, 20
Aix galericulata, 9, 20, 34
Aix sponsa, 9, 20, 34
Alien species, introduction of. *See* Introduced
      species
The American Ornithology (Wilson), 64
Artistic works
   American, *54, 60–63, 189–191, 195*
   Chinese, 175–176, *176, 178,* 179–181,
      182
   European, *56,* 176–178, *178, 179*
   hunting/postage stamps, 189–190,
      189–191, 196
   Japanese, *174, 177,* 181–185, *182–185*
   literature, 179–180
   poetry, 174–175, 183–185
   Russian, *196*
   songs, 175, 179–180
Asia. *See* Far East
Asian flyway, 115, *116,* 117–118, 121, 134,
      198
   See also Far East; Flyways
Atlantic Flyway, 19, 85, 90, 93
Audubon, John James, *54,* 55–56, 65, 68–70,
      72–73, 105
*The Audubon Society Field Guide to North
   American Birds* (Udvardy), 117

Banding
   Bird Banding Laboratory, 221
   Great Britain, *162,* 171–172, *171*
   Japan, 148, 150
   need for, in Far East, 118, 150, 171, 223
   nestbox programs and, *96, 211,* 214–215,
      221–223

North America, 214–215, 221–223
   number banded, per year, 221
   timing of, *96,* 221
   See also Traps and trapping programs
Behavior
   decoying, *48*
   feigning, *48,* 127
   fighting, *51,* 52
   foraging, 48, 89, 130
   homing instinct, 45, *52,* 94–95, 125
   nesting, 45, *46–49,* 60
   nocturnal, 113, 147, 169
   perching, *94*
   posturing, *44,* 45
   preening, *51, 88,* 89, 176
   protective, 45–48, *46–49, 125–127,* 127,
      188
   roosting, *111*
   secretive, 34, 74, 84–85, 110, 152–153,
      157, 161, 173
   temperamental differences in, 52, 103,
      163–164
   See also Breeding; Comparisons; Courting
      behavior; Migration
Bell, David, 120, 203–204
Bellrose, Frank, 11, 56, 75, 80, 111
   conservation and, *194, 195,* 217
Bent, Arthur Cleveland, 83, 109–110
Birds of America (Audubon), 55
Birds of Asia (Gould), 178
The Birds of Ussuriland (Knystautas &
      Shibnev), 132–133
Borcharnikov, Vladimir N., 133, 203–204,
      203
Brazil, Mark, 154–155, 157
Breeding
   broods per season, 112
   clutch size, 76, 100, 112–113, 126, 170
   crossbreeding, 10, 45, 206–208
   diet, 100–102, 125
   drake and, 109–110, 125, *125*
   egg-laying, 34, 99–102, 125–126, 169

hatching, *30,* 102–106, *102*
   human presence and, 102–103
   hybrids, 45, 206–208
   imprinting, 25, 103–105, 107
   incubation period, 34, 102–104, 125
   infertility, 45, 112
   predators during, 102
   renesting, 76
   See also Courting behavior; Eggs; Habitat;
      Nests
British and Irish List, 160, 167
Broods and brooding
   behavior, 109–113, 125–127, *125–127*
   diet during, 107–109, 126–127
   habitat, ideal, *31, 38,* 111
   mixed broods, *24, 42–45, 43, 53,* 163
   number per season, 112
   survival rates, 106, 109–112, 126
   See also Breeding; Ducklings; Habitat

California
   central, 10, 19, 49, 52, 91, 112, 194, 220
   northern, *9–13, 15–53, 73, 77,* 108
   See also Indian Meadow Ranch
Canada, 59, 75, 90, 91, 93
Central Flyway, 85
Chicks. *See* Ducklings
China
   conservation, 136–139, 143–145,
      198–200
   habitat, 115, 131, 134–140, 170
   historical accounts, 134, 164
   hunting, 140–141, 144, 148
   Mandarin's decline in, 7, 37, 166–167
   migration, 131, 134–138
   nature reserves, 136–139, 143–145
   population, 139
   range, *18,* 115–118, *116,* 134–139, *135*
   See also Artistic works; Far East
Chinese (Scaly-sided) Merganser, 120, 129,
      137–138, 203–205, *204*
Classification. *See* Mandarin; Wood Duck

Clutches
  second, 106, 112–113
  size, 76, 100, 126, 170
  See also Breeding; Eggs; Nests
Coloring *See* Plumage
Commercial hunting. *See* Hunting
Comparisons
  behavior, *44*, *45*, *46*, *47*, 49, *49*, 127
  chromosomal count, 45
  clutch size, 76, 100, 126, 170
  courtship, 49, 52
  diet, 45–46, 48, 129–130
  drakes, 20–22, 49, 52
  ducklings, 22–24, *24*, 126–127, 163, 169
  eggs, 48, 125–126, 163
  feeding, 45–46, 48
  hens, 22–23, *22–23*, 46–47, 49, *49*, 52
  migratory instinct, 18, 48–49, 52–53,
    164–165
  native habitats, 15, 38, 42, *77*, 87,
    115–118, 186
  nesting, *25*, 45, 48, 169
  pair bonding, 47, *47*
  plumage, 20–24, *22–24*, 97
  size, 48, 163
  temperament, 52, 103, 163–164
  voice, *25*, 48–49
  water-type preferences, 122
  weight, 163
  See also Species/genus interrelationship
Compatibility, 12, *14*, 15–53, *16–17*, *40–42*,
    50–53
  See also Species/genus interrelationship
Conservation, 12, 37, 123, 167
  Chinese (Scaly-sided) Merganser, 120,
    129, 137–138, 203–205, *204*
  hunting stamp programs, 157, 189–190,
    189–191
  introduced species and, 200–208
  Mandarin, 118, 129, 136–145, 148,
    196–213
  Nene Goose, 201–204
  North America, 188–196
  postage stamps, *190*, *196*
  White-headed Duck, 206–208, *207*,
    210–211
  White-winged Wood Duck, 34, 203–204,
    211
  Wood Duck, 70–81, 91, 188–196
  See also Hunting; Laws and legislation;
    Nature reserves
Conservation organizations
  Asian Wetlands Bureau, *209*

Birdlife International, *209*
California Waterfowl Association, 214–215
Ducks Unlimited, 80, *191*, 191–192,
  210–211, 219
International Crane Foundation, 199,
  210–211
International Waterfowl and Wetlands
  Research Bureau, 203–208, *207*, *209*
Lake Baikal Watch, 199
National Audubon Society, 191–192, 195,
  199, *209*, 210–211
National Geographic Society, 210
Nature Conservancy, 191–192, *209*, 210
Oshidori Trust, 212
Siberian Forest Protection Project, 199
Sierra Club, 191, 210
Socio-Ecological Union, 199
Species Survival Commission, *209*, 210
Threatened Waterfowl Research Group,
  209, 210
Waterbird Specialist Group, *209*
Wild Bird Society of Japan, 152, 154, 199,
  210
Wilderness Society, 199
Wildfowl and Wetlands Trust, 120, 163,
  169, 191, 201–210, *209*
World Conservation Monitoring Centre,
  209
World Conservation Union, 206, 208, 210
  organizational chart, *209*
World Pheasant Association International,
  210–211
World Wide Fund for Nature, *209*, 210
World Wildlife Fund, 191, 210
  See also Nature reserves
Courting behavior
  displays, *44*, 45, *158–159*, 176
  drakes, 49, 52, *69*
  hens, 49, 52
  rituals, *88*, 89, 92–93, 123–124, *123*, *132*
  voices, 48
  See also Breeding; Mating; Pairing behavior
Crossbreeding, 10, 45, 206–208

Davies, Andy, 11, 153, 167–173, *171*
Delacour, Jean, 34, 164–165
Dermid, Jack, 107–108, 111
Diet
  breeding, 100–102, 125
  differences in, 129–130
  duckling, *31*, 107–109, 126–127, 130, *130*
  favored, 26, 29, *38*, 89, *89*, 100, 122
  range of foodstuffs in, 77

  See also Feeding; Food sources
Die Vogelwelt Ussuriens (Knystautas &
    Shibnev), 117
Differences, between Mandarin, Wood Duck.
    See Comparisons
Displays, courting. *See* Courting behavior
Distribution. *See* Flyways; Population; Range
Drop nests, 100, *100*, 170–171
Ducklings
  adoption, 103–107, *106*, 112
  diet, *31*, 107–109, 126–127, 130, *130*
  hatching, *30*, 102–106, *102*
  leaving the nest, *30–31*, 43, 61, 65,
    104–109, *104–107*, 126, *164*
  mixed broods, *24*, 42–45, *43*, 53, 163
  mortality rate, 109
  number of, compared, 126
  plumage, 22–23, *24*
  precocial trait, 105, *107*
  predators, 109–112
  survival rates, 106, 109–112, 126
  web-tagging, 214, 221
  See also Breeding; Broods and brooding;
    Eggs; Nests
Dump nests, 100, *100*, 168

Eclipse molt, 23, 109–110, 128
Ecology and Management of the Wood Duck
    (Bellrose & Holm), *194*, 194
Eggs
  abandoned, *34*, 100
  adopted, 103–107, 112
  color, 48, 125–126
  drop nest, 100, *100*, 170–171
  dump nest, 100, *100*, 168
  egg tooth, *102*, 103
  hatching, *30*, 102–106, *102*
  incubation period, 34, 102–104, 125
  infertile, 45, 112
  laying, 34, 99–102, 125–126, 169
  membrane, 35
  number of. *See* Clutches
  pipping, 47, 103, 105
  predators. *See* Predators
  production rates, 100, 126
  time of year first noted, 96
  See also Breeding; Nests
Endangered species. *See* Conservation;
    Mandarin
England. *See* Great Britain
Environmental Agency of Japan, *151*, 152, 154
Environmental protection. *See* Conservation
Exodus of ducklings, from nest. *See* Ducklings

Extinction, threats of. *See* Conservation

Fallodon Papers, 162
Far East
  endangered Mandarin in, 12, 37, 123, 167
  flyways, compared to Pacific Flyway, 38, 115–118
  habitat, 37, 115–118, 121, 128–129
  hunting, 118, 129, 140–141
  See also China; Japan; The Koreas; Siberia; Taiwan; Ussuriland
Feathers. *See* Plumage
Feeding
  during breeding, 100–101, 125
  foraging behavior, 48, 89, 130
  nocturnal, 169
  programs, for waterfowl, 37
  times, compared, 45–46
  See also Diet; Food sources
Feral Mandarin, introduction of. *See* Introduced species
Fertility rates, 45, 112
Flightless period, 23–24, 110, 128
Flyways
  Asian, 115, *116,* 117–118, 121, 134, 198
  Atlantic, 19, 85, 90, 93
  Central, 85
  latitude and, 38, 169
  Mississippi, 9, 19, 85, 90, 93
  Pacific. *See* Pacific Flyway
  similarities in, 38, 115–118
  See also Migration; Range
Folklore
  on ducklings leaving the nest, 61, 65, 105
  on Mandarin's cleverness, 163–165
  on monogamous pair bonds, 175–180, 185
Food sources
  Far East, 121–122, 128–130, *137,* 147–148, 171
  North America, 26–27, 29, 31–32, *38,* 88–89, 92, 100–102, 111
  See also Diet; Feeding
Freshwater wetlands. *See* Habitat
Fujimaki, Yuzo, 154

Gould, John, 177–178
Great Britain
  conservation, 160–173, 202–210
  habitat, *161,* 164–165, *168,* 170
  introduced Mandarin, 12, 160, 164–165, 169, 177
  private collections in, 11, 63, 160–166
  range, *161*

trapping programs, *162,* 171–173, *171*
Wood Duck in, 153, 163–164, 169
Grey, Lord, of Fallodon, 63, 161–164, *163*

Habitat
  attractive to Mandarin, Wood Duck, 37, *38, 42, 77,* 87, *186*
  essential characteristics, 29, 87, 92
  Far Eastern, compared to Pacific Flyway, 38, 115–118
  human presence in, 24, 84, 98–99, 102–103, 150–152, 220
  nestboxes, impact on, 80–81
  pioneering, 76–77, *76,* 84, 139, 170
  ponds, 37, 76–77, *76, 77*
  terminology, 86
  threats to, 73–75, 77–78, 121, 128–129, *192, 198*
  usable, *18–19*
  wetlands as, 24, 27, 29, 76, 84–86, 92, *92,* 155–156, *186,* 187
  See also Adaptation; Conservation; Indian Meadow Ranch; Nests
Hatching, *30,* 102–106, *102*
  See also Breeding; Eggs
Hawkins, Arthur, 11, 56, 194, 215
Hester, Eugene, 107–108, 111
Holm, Daniel J., 194
Homing instinct, 45, *52,* 94–95, 125
Hughes, Barry, 120, 133, 203–205
Hunting
  commercial, 55, 71–74, 85, 188–189, 196
  Far East, 118, 129, 130, 140–141, 196
  Great Britain, 167
  impact of, 70–76, 91, 92
  legislation, 75, 78–79, 189–190
  North America, 57, 59, 60, 70–81, 73, 85
  population estimates and, 91, 157
  stamp programs, 157, 189–190, *189–191*
  See also Banding; Conservation; Laws and legislation
Hybrids, 45, 206–208

Imprinting, 25, 103–105, 107
  See also Breeding
Incubation period, 34, 102–104, 125
Indian Meadow Ranch, California, 15, *38, 40, 195*
  ecology, 25–27, 108
  location, 19–20, *19,* 25
  Mandarin program, history of, 15–53, *35*
  nestboxes, 27, 29–31, *29,* 218–220, *218–220*

predators, 30–32, *32, 101,* 103, 111
wildlife, 26, 27, *28,* 29–30, 35–36, 39, *42, 69,* 77, 95
Wood Duck program, history of, 27–34
See also California; Mandarin; Wood Duck
Interrelationships. *See* Comparisons; Compatibility; Species/genus interrelationship
Introduced species
  conservation and, 200–208
  Mandarin as
    to Great Britain, 12, 160, 164–165, 169, 177
    to North America, 9–13, 18, 34–36
  risks of, 206–208
  Ruddy Duck as, 206–208, *207*
  See also Compatibility; Species/genus inter-relationship

Japan
  conservation, 148, 198, 212–213
  habitat, 38, 115–118, 131, 137, 146–152, 155–156
  migration, 137–138, 147
  population, 149, 152–155, 166–167
    midwinter counts 1973–1995 (chart), *151*
  range, *18,* 115–118, *116, 147*
  See also Artistic works; Far East

Kear, Janet, 11, 169, 201–202
Knystautas, Algirdas, 117, 132–133
The Koreas, 18, 115, 119, 131, 137–138, 147

Laws and legislation
  Migratory Bird Hunting Stamp Act, 189–190
  Migratory Bird Treaty Act, 75, 78–79, 189
  Pittman-Robertson Act, 190
  Weeks-McLean Bill, 189
  Wildlife Restoration Act, 190
  See also Conservation
Legends, about Mandarin, Wood Duck. *See* Folklore
Leopold, A. Starker, 9–11, 208
Leopold, Aldo, 9, 80, 108, 187, 196
Leopold, Frederic, 9, 11, 108–109, *194,* 215
Lever, Sir Christopher, 11, 169, 202
Lord Grey of Fallodon. *See* Grey, Lord, of Fallodon

Mallard, 39, 45, 55, *63,* 77, 81
  compared to Wood Duck, 20, 106–107
  nests of, *93*

Man and Wildfowl (Kear), 169, 201–202
Manchuria, 119, 131, 136, 140
Mandarin
  compatibility, with Wood Duck, 12, *14,*
    15–53, *16–17, 40–42, 50–53*
  earliest references to, 178
  endangered, 12, 37, 123, 167
  folklore about. *See* Folklore
  at Indian Meadow Ranch, 9–13, 15–53
  names for
    common, 11, 15, 115, 134, 176–177,
     179
    scientific, 9, 20, 34
  population of. *See* Population
  taming of, 161–165
  See also Artistic works; Comparisons;
    Conservation; Far East; Great Britain;
    Indian Meadow Ranch; Introduced
    species; *specific topics*
Man-made nestboxes. *See* Nestboxes
Maps
  China, *18, 116, 135*
  Far East, *18, 116*
  Great Britain, *161*
  Japan, *18, 116, 147*
  North America, *19, 85*
  Ussuriland, *18, 116, 119*
The Market Book (Reeves), 71–72
Mating, 89, *95,* 123–124, *123*
  See also Breeding; Courting behavior
Measurements
  compared to Mallard, 20
  duckling, 23
  eyes, 24, *33,* 95, *124*
  mature duck, 48
  tail length, 95
  weight, 163
  wings, 24, *33,* 86–87, 95, *133, 136*
Merganser, Chinese (Scaly-sided). *See* Chinese
  (Scaly-sided) Merganser
Migration
  homing instinct, 45, *52,* 94–95, 125
  hunting season and, 70–76, 91, 92
  Mandarin
    Far East, 115–119, 123–124, 130–131,
     135–138
    flight speed of, 147
    instinct for, 18, 49, 164
  Wood Duck
    fall, 52–53, 89–93, 113, 130–131
    flyways of. *See* Flyways
    outcrossing and, 93
    resident *versus* migratory, 19, 76, 85,

     91–92
    spring, 93–96, 111–112
    See also Flyways; Nests; Population; Range
Mississippi Flyway, 9, 19, 85, 90, 93
Missouri River Journals (Audubon), 56
Molts and molting
  eclipse, 23, 109
  remigial, 23–24, 110
  spring, 23–24
  summer, 22, 89, *108,* 109, *110,* 128
  See also Plumage
Monogamous pair bonds, 175–180, 185
Murofushi, Yuzo, 148–151, 154, 212
Myths. *See* Folklore

Names, for Mandarin, Wood Duck. *See*
  Mandarin; Wood Duck
Natural cavities, as nests. *See* Nests
Naturalized Animals of the British Isles
  (Lever), 169
Naturalized species. *See* Introduced species
Nature reserves
  in China, 136–139, 143–145
  in Russia, 129
  in the United States, 191–193, *193*
  See also Conservation
Nestboxes
  building, 27, 214–220
  entry holes, *153,* 215, 219
  human presence and, 24, 98–99, 150–152,
    220
  importance of, to research, 98, 168,
    214–215
  location of, 29, 87, 92, 215
  maintenance, 31–32, *34,* 96, 220
  Mandarin's preference for, 25, 48
  need for, 80–81, 98, 168, 214–215
  number of, in use, 80, 98, 214–215
  photographs, 25, 29, 35, *46,* 52, 79, 96–98,
    104, 105, 211, 214, 216, 218–219
  predator deterrence, 215, 219–220, *220*
  programs of, 9–10, 112, 194–195,
    214–215
  recordkeeping, 31–32, 215, 220
  role of, in species survival, 80–81, 98,
    214–215
  See also Nests; Traps and trapping programs
Nesthouses. *See* Nestboxes
Nests
  abandoned, 100
  competitors for, 94–95, *94*
  drop, 100, *100,* 170–171
  ducklings leaving. *See* Ducklings

  dump, 100, *100,* 168
  ground-nesting ducks, compared, *93,*
    94–95, 168
  man-made. *See* Nestboxes
  search for, 96–99, *98, 99,* 124–125
  second, 76, 106, 112–113
  tree cavities as
    depth, 95
    essential traits, 39, 97, 121, 124–125
    homing instinct, 94–95, 125
    photographs, *25, 30, 59, 66–67, 93, 124,*
     156, 214
    usage of, *versus* nestboxes, 97–98
  See also Breeding; Habitat; Nestboxes
Nest sites
  best, for brooding, 107–109
  Mandarin, compared to Wood Duck, 45
  man-made. *See* Nestboxes
  near human habitation, 24, 34, 84
  selection process, 39, 96–98, *98, 99*
  tree cavities as. *See* Nests
  See also Adaptation; Breeding; Habitat;
    Nestboxes
Netherlands, 164–165, 208
North American Wood Duck. *See* Wood Duck
Northern Wood Duck. *See* Wood Duck
North Korea. *See* The Koreas
Nutrition. *See* Diet

Organizations, conservation. *See*
  Conservation organizations
Ornithological Biography (Audubon), 68–70

Pacific Flyway
  geography, 25–27, 38, 85
  range, 15, *19*
  similarity with Asian flyway, 38, 115–118
  See also Flyways; Migration; Range
Pairing behavior, *37,* 47–48, *47,* 89, 95–96,
  95
  monogamous, 175–180, 185
  See also Breeding; Courting behavior;
    Mating
Philopatry. *See* Homing instinct
Pioneering
  of new habitat, 76–77, *76,* 84, 139, 170
  See also Adaptation
Pipping, 47, 103, 105
  See also Hatching
Plumage
  crest, 21–22, 49, *140*
  drakes, 20–23, *20, 21, 36, 140, 166, 167*
  eclipse, 110

Plumage *(continued)*
  eye, 22, *22*, 23, *60*, 95
  flight feathers, 21–24, 89, *97*, *108*, 110,
    128, *167*
  flightless period, 23–24, 110, 128
  hens, *22–23*, 24, 49, *64*, *65*, *167*
  historical accounts, 58–62, 64–65, 68–70
  juvenile duck, *109*
  mixed broods, *24*
  nuptial, 52, 89, 92, *110*, 128, *167*
  sail feathers, 9, 21, *21*, 23, 128, *131*,
    158–159, *166*, *167*
  tail, 20
  tertial, 21
  underfeathers, 22
  See also Molts and molting
Poaching, 129, 144
  See also Predators
Ponds, as habitat, 37, 76–77, *76*, *77*
Population
  Mandarin
    China, 139
    Far East, 117–118, 132–134
    Great Britain, 160, 171–173
    Japan, 149, 151–156, *151*
    North America, 9, 18, 37
    Ussuriland, 132–133
  Wood Duck
    Great Britain, 153, 163–164, 169
    North America
      current, 18–19, 76–81, 84, 188–190
      historical, 55–76
      hunting and, 70–76, 91, 92
    See also Flyways; Migration; Range;
      Survival rates
Postage stamps, 190, *196*
Predators
  in Far East, 128, *153*, *156–157*
  in North America, 30–32, *32*, 92, *101*,
    102–103, 109–112, 215, 218–220,
    220
  See also Poaching

Range
  Mandarin
    China, *18*, 134–139, *135*
    Far East, *18*, 115–118, *116*
    Great Britain, *161*
    Japan, *18*, *147*
    North America, 18–19, *19*, 84–85, *85*
    Ussuriland, *18*, *119*
  Wood Duck, 15, 18–20, *19*, 84–85, *85*
  See also Flyways; Migration; Population

The Recreation (Lord Grey of Fallodon), 63
Red Data Books, *209*, 210
Red Lists, *209*
Reeves, H. Milton, 11, 56–58, 71–72
Reserves, nature. *See* Nature reserves
Russia. *See* Ussuriland

Sail feathers. *See* Plumage
Scaly-sided (Chinese) Merganser. *See*
    Chinese (Scaly-sided) Merganser
Scientific names, for Mandarin, Wood Duck,
    9, 20, 34
Scott, Sir Peter, 11, 166, 187, 202, *210*
Self-introduced species. *See* Introduced
    species
Shibnev, Yuri, 117, 132–133, *203*
Siberia, 115, 119–120, *120*, 147, 202
Similarities, in Mandarin, Wood Duck. *See*
    Comparisons
Size. *See* Measurements
Sounds, of Mandarin, Wood Duck. *See* Voice
South Korea. *See* The Koreas
Species/genus interrelationship, 9, 15, 20,
    59, 83–84
  See also Comparisons; Compatibility
Stamps
  California duck stamp, 190, *190*
  federal duck hunting, 157, 189–190,
    189–191
  postage, 190, *196*
Survival rates, 106, 109–113, 126
  See also Conservation; Population

Taiwan, 18, 115, 140, 147
Territory. *See* Flyways; Range
*Thirty-five Year Study of Wood Ducks on the
    Mississippi* (Leopold, Frederic), 194
Traps and trapping programs
  box traps, 221–222
  Far East, need for, 148, 150, 223
  Great Britain, 171–173, *171*
  North America, 221–223, *221–223*
  timing of, 221
  Y-traps, 222–223, *222–223*
  See also Banding; Nestboxes
Tree cavities, as nests. *See* Nests

Udvardy, Miklos D. F., 117
United Kingdom. *See* Great Britain
U.S. Fish and Wildlife Service, 76, 157, 192,
    214–215, 221
Ussuriland
  fall migration, 130–132, 136–137,

146–147
  food sources in, *128*, 129–130, *129*
  habitat, 119–123, 120–121, *120–121*,
    205
  nesting grounds, 121–127
  population, 132–133
  predators, 128–129
  range, *18*, 115–118, *116*, 134
  spring migration, 120, 123–124
  See also Artistic works; Far East

Voice
  alarm calls, 39, 49, 86
  compared, 25, 37, 48–49
  drakes, 37, 48
  hens, 49, 103
  imprinting and, 103
  warning, 49

Waterfowl
  family Anatidae, 83–84
  See also Conservation; Introduced species;
    Nature reserves
Web-tagging, of ducklings, 214, 221
Wetlands. *See* Habitat
Wild California (Leopold, A. Starker), 11
Wildfowl. *See* Waterfowl
Wildlife sanctuaries. *See* Nature reserves
Wilson, Alexander, 63–65, 68
Wood Duck
  compatibility with Mandarin, 12, *14*,
    15–53, *16–17*, *40–42*, *50–53*
  earliest reference to, 58
  first use of term, 62–63
  folklore. *See* Folklore
  names for
    common, 27, 58–59, 60–62, 83, 86
    scientific, 9, 20, 34
  population. *See* Population
  symposiums, 57–58, 194
  See also Artistic works; Behavior;
    Comparisons; Great Britain; Indian
    Meadow Ranch; Species/genus
    interrelationship; *specific topics*

ACKNOWLEDGMENTS

Lawton L. Shurtleff and Christopher Savage would like to thank the following:

Sir Peter Scott and Lady Scott, for their help and inspiration;

Dr. Janet Kear and Sir Christopher Lever, for years of unstinting advice and encouragement and for reviewing the manuscript;

Andy Davies, for sharing his pioneering research and findings on the Mandarin in England;

Dr. Myrfyn Owen, director general of the Wildfowl and Wetlands Trust, and the WWT staff members, and Dr. Mike Moser, for sharing their research and publications and for their personal participation;

Dr. Barry Hughes and Dr. David Bell, for their observations in the Bikin River Valley of Ussuriland;

Dr. Vladimir Borcharnikov, for his observations in Russia;

Dr. Hiroyoshi Higuchi, director of the Wild Bird Society of Japan, Professor Yuzo Fujimaki, Yuzo Murofushi, Dr. N. Kuroda, Dr. Mark Brazil, and Ichiro Kikuta, for their observations in Japan;

Jeffrey Boswell of the BBC, Mark Beaman, Professor Cheng Tso-hsin, Professor Zhao Zhengjie, and Dr. Jianjian Lu, for their observations in China;

Dr. Lucia Liu Severinghaus, for her observations in Taiwan;

Professor Won Pyong-Oh and Duncan Poore, for their observations in North and South Korea;

Dr. G. H. Voorwijk, for his observations in the Netherlands;

Tim Sullivan and Elizabeth McCance, assistants-to-the-chair of the Species Survival Commission of the World Conservation Union, for a glimpse inside the workings of the world's largest conservation organization;

and the many, many other individuals who helped along the way.

Lawton L. Shurtleff would like to thank the following:

Frank C. Bellrose, the dean of Wood Duck scholars, who, while working on his monumental *Ecology and Management of the Wood Duck*, shared his extraordinary knowledge, reviewed the entire manuscript, and even traveled to observe firsthand the Wood Ducks and Mandarin at Indian Meadow Ranch in California;

Henry M. Reeves, for sharing with me his enthusiasm for the Wood Duck and his knowledge of its history;

Dr. A. Starker Leopold, for his early encouragement;

Dr. S. Dillon Ripley, Frederic Leopold, Art Hawkins, Jack Dermid, and Dr. Miklos D. F. Udvardy, for their advice and expertise;

Richard Cuneo and Kendrick Morrish, for their friendship and guidance;

Carl Sams II, Jean Stoick, and Greg Nelson, who came to Indian Meadow Ranch and made what are unquestionably the finest photographs of free-flying Mandarin ever produced, and Carl in particular, for introducing us to the art of waterfowl photography and thus enabling us to practice it on our own, and for giving generously of his time and considerable talent throughout the making of this book;

Judith Dunham, our editor and polisher of prose for all these years, and without whose insightful guidance there never would have been a book;

George Ackerman and Clyde Rash, for assisting in the hard work of maintaining, studying, and photographing the Wood Duck and the Mandarin at Indian Meadow Ranch;

Kathy Mayer, Arlene Carramusa, Robyn Garcia, and Carolyn Klinepeter, for learning to read my scribbling and turning it magically into type;

Bobbie and Andrew, who were there at the beginning;

My daughter, Lynda, and my two sons, Bill and Jeffrey, one an author and the other a purveyor of books, for their valuable insights into the making of a book;

Anneke Shurtleff, for her patience and understanding during ten years of work on this book and for her astute and perceptive reading of the manuscript and review of the photographs at every stage.

Christopher Savage would like to thank the following:

His Royal Highness Prince Hitachi, for his encouragement;

Yuzo Murofushi and Miyauchi Kazunori, founders with me of the Oshidori Foundation;

Dr. George Archibald, for contributing his experiences in crane conservation in the former U.S.S.R.;

Dr. Duncan Parish of the Asian Wetland Bureau;

Dr. Yoshimitsu Shigeta of the Yamashina Institute;

Moto Tatsumura, for his help in contacts with the Shosoin at Nara;

Professor Vladimir Flint, for his encouragement and guidance;

Ruth Sloan, Marina Tchelitchev, and Simon Reid, for their help in the translations from the Russian;

Anne Savage, for tolerating my prolonged absences in the Far East and for her constant support and enthusiasm.

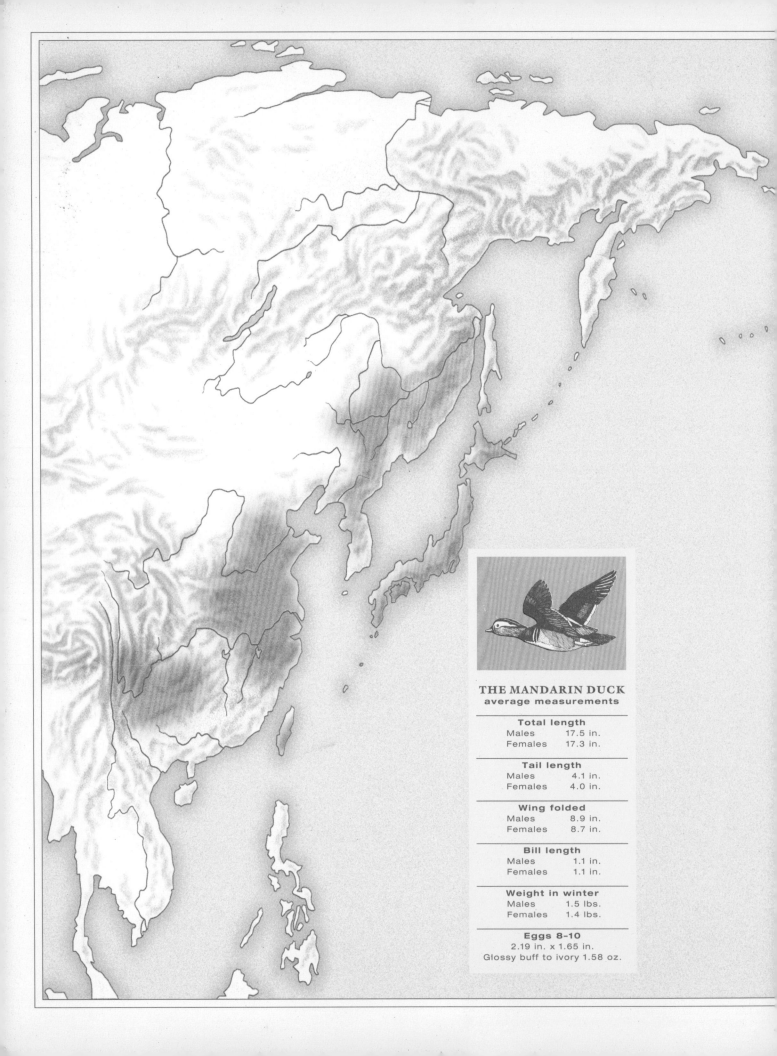

THE MANDARIN DUCK
average measurements

Total length
Males   17.5 in.
Females   17.3 in.

Tail length
Males   4.1 in.
Females   4.0 in.

Wing folded
Males   8.9 in.
Females   8.7 in.

Bill length
Males   1.1 in.
Females   1.1 in.

Weight in winter
Males   1.5 lbs.
Females   1.4 lbs.

Eggs 8-10
2.19 in. x 1.65 in.
Glossy buff to ivory 1.58 oz.